Hazardous Air Pollutants

Case Studies from Asia

Hazardous Air Pollutants

Case Studies from Asia

Edited by
Dong-Chun Shin

CRC Press
Taylor & Francis Group
Boca Raton London New York

CRC Press is an imprint of the
Taylor & Francis Group, an **informa** business

CRC Press
Taylor & Francis Group
6000 Broken Sound Parkway NW, Suite 300
Boca Raton, FL 33487-2742

Printed on acid-free paper
Version Date: 20160104

International Standard Book Number-13: 978-1-4665-9356-5 (Hardback)

Library of Congress Cataloging-in-Publication Data

Names: Shin, Dong-Chun, editor.
Title: Hazardous air pollutants : case studies from Asia / edited by Dong-Chun Shin.
Other titles: Hazardous air pollutants (Boca Raton, Fla.)
Description: Boca Raton : Taylor & Francis, CRC Press, 2016. | Includes bibliographical references and index.
Identifiers: LCCN 2015048517 | ISBN 9781466593565 (hardcover : alk. paper)
Subjects: LCSH: Air--Pollution--Health aspects--Asia--Case studies. | Air--Pollution--Asia--Case studies. | Air quality management--Asia--Case studies.
Classification: LCC RA576.7.A78 H39 2016 | DDC 363.739/2095--dc23
LC record available at http://lccn.loc.gov/2015048517

Visit the Taylor & Francis Web site at
http://www.taylorandfrancis.com

and the CRC Press Web site at
http://www.crcpress.com

Contents

Preface

Most Asian countries took their first steps later than did Western countries in terms of industrialization, but they are experiencing advancements much faster. Accordingly, the numbers of power plants and vehicles dramatically increased without a sufficient consideration for shadows of industrialized society. During the early period of industrialization in the Republic of Korea, the only thing we focused on was water quality because it was easy to determine how muddy water was. However, after our economic development, we started to recognize how cloudy our skies were, and simultaneously, or maybe earlier, a substantial number of people began to die of the effects of air pollution. At that time, almost all people were not aware that the air, an essential element of our lives, could contain hazardous particles and cause diseases and even deaths.

According to the World Health Organization, approximately 7 million people worldwide die each year because of exposure to indoor and outdoor air pollution. Rapid urbanization, tremendous population growth, growth, and economic development seem to have worsened air quality especially in the largest Asian cities, such as Beijing and New Delhi, which is worth paying attention to in order to improve public health in Asia. In addition to, the patterns of morbidity and mortality in low-income Asian countries suggest these are in a transition period because of increasing life expectancy as well as greater prevalence of risk factors such as smoking, drinking, and sedentary lifestyles. More specifically, the number of patients with chronic illnesses, such as cardiopulmonary diseases, who are susceptible to air pollution has been rapidly increased for decades.

This book concentrates on linking air pollution to health in Asia and includes six case studies: China, Japan, Taiwan, Indonesia, Malaysia, and the Republic of Korea. Prestigious researchers were invited to contribute to this important work, encompassing the cardiovascular system, respiratory system, reproductive system, and mortality linked to air pollution in Asia. In this book, Section I summarizes demographics, sources of air pollution, epidemiological findings, biological mechanisms, exposure assessment, and risk assessment for each country. Section II introduces new fields of air pollution research such as air pollution effects on the brain, the impact of vehicle emission regulations, and climate change. I thank all the contributors for their time and efforts to complete these chapters.

Through this book, we aim to offer an opportunity for future researchers to share research experiences on air pollution and health. For readers, this book provides a glance at up-to-date air pollution levels, health findings, and air quality policies of Asian countries. On behalf of all the contributors, I hope this book will extend the readers' understanding about air pollution effects on public health in Asia and inspire future researchers.

Dong-Chun Shin

Editor

Dong-Chun Shin, MD, PhD

Education

- MD, Yonsei University College of Medicine (1980)
- PhD, Graduate School of Public Health, Yonsei University (1989)
- VS, Department of Environmental and Industrial Health, School of Public Health, University of Michigan, Ann Arbor, MI (1991–1993)

Present Position

- Professor/Director, Department of Preventive Medicine/Institute for Environmental Research, Yonsei University College of Medicine
- President, Korea Society of Risk Governance (KOSRIG)
- Chair, Society for Risk Analysis (SRA)—Korea Chapter
- Board of Korea Society of Preventive Medicine
- President, Korea Society for Green Hospitals
- Chair, Policy Studies Division, Korea National Academy of Science and Technology
- Environmental Health Committee, Ministry of Environment
- Council Member, World Medical Association (WMA)
- Chair, Environmental Caucus, World Medical Association (WMA)
- Council Chair, Confederation of Medical Associations in Asia and Oceania (CMAAO)

Past Experience

- President, Korean Society for Indoor Environment
- President, Korean Society of Environmental Toxicology
- Organizing Chair, Joint Conference of ISEE–ISES (International Society for Environmental Epidemiology–International Society for Exposure Sciences)
- Vice President, Korean Society for Atmospheric Environment
- Central Advisory Committee, Ministry of Environment
- Vice Dean, Yonsei University College of Medicine
- Vice Director, Department of Planning and Coordination, Yonsei University Health System
- Director, Department of Administration, Yonsei University Health System

Society Membership

- Dioxin and Persistent Organic Pollutants Conference
- International Society of Exposure Science

- International Society of Indoor Air Quality and Climate
- Korean Society for Atmospheric Environment
- Korean Society for Indoor Environment
- Korean Society of Environmental Health and Toxicology
- Korean Society of Occupational Medicine
- Korean Society of Preventive Medicine
- Society for Risk Analysis

Awards

- 2004 Presidential Citation for Environmental Achievement, Korea
- 2006 Order of National Service Merit, Korea

Research Area

- Risk Assessment and Management, Health and Environmental Sustainability

Publications

- 158 peer-reviewed articles, including 36 SCI-listed

Contributors

Amir Afiq Abdullah
Department of Environment
Ministry of Natural Resources and Environment
Putrajaya, Malaysia

Umar-Fahmi Achmadi
Department of Environmental Health
University of Indonesia
Depok, Indonesia

Susan Buchanan
School of Public Health
University of Illinois at Chicago
Chicago, Illinois

Erica Burt
School of Public Health
University of Illinois at Chicago
Chicago, Illinois

Chang-Chuan Chan
College of Public Health, Institute of Occupational Medicine and Industrial Hygiene
National Taiwan University
Taipei City, Taiwan

Szu-Ying Chen
College of Public Health, Institute of Occupational Medicine and Industrial Hygiene
National Taiwan University
Taipei City, Taiwan

Chia-Pin Chio
College of Public Health, Institute of Occupational Medicine and Industrial Hygiene
National Taiwan University
Taipei City, Taiwan

Jaelim Cho
Department of Occupational and Environmental Medicine
Gachon University Gil Medical Center
Incheon, Korea

Er Ah Choy
Faculty of Social Sciences and Humanities
Universiti Kebangsaan Malaysia
Bangi, Malaysia

Xinbiao Guo
School and Department of Occupational and Environmental Health Sciences
Peking University School of Public Health
Beijing, China

Md Firoz Khan
Research Center for Tropical Climate Change System, Institute of Climate Change
Universiti Kebangsaan Malaysia
Bangi, Malaysia

Changsoo Kim
Department of Preventive Medicine
Yonsei University College of Medicine
Seoul, Korea

Mohd Talib Latif
Faculty of Science and Technology, School of Environmental and Natural Resource Sciences
Universiti Kebangsaan Malaysia
Bangi, Malaysia

Norhayati Mohd Tahir
Environmental Research Group, School of Marine Science and Environment
Universiti Malaysia Terengganu
Terengganu, Malaysia

Chris Fook Sheng Ng
Department of Human Ecology, School of International Health, Graduate School of Medicine
University of Tokyo
Tokyo, Japan

Peter Orris
School of Public Health
University of Illinois at Chicago
Chicago, Illinois

Rachmadhi Purwana
Department of Environmental Health
University of Indonesia
Depok, Indonesia

Mazrura Sahani
Environmental Health and Industrial Safety Program, School of Diagnostic Science and Applied Health, Faculty of Health Sciences
Universiti Kebangsaan Malaysia
Bangi, Malaysia

Eri Saikawa
Department of Environmental Studies
Emory University
Atlanta, Georgia

Noelle E. Selin
Institute for Data, Systems and Society and Department of Earth, Atmospheric and Planetary Sciences
Massachusetts Institute of Technology
Cambridge, Massachusetts

Dong-Chun Shin
Department of Preventive Medicine
Yonsei University College of Medicine
Seoul, Korea

Toru Takebayashi
Department of Preventive Medicine and Public Health
Keio University School of Medicine
Tokyo, Japan

Wan Rozita Wan Mahiyuddin
Institute for Medical Research, Ministry of Health
Kuala Lumpur, Malaysia

Shaowei Wu
Department of Environmental Health
Harvard T.H. Chan School of Public Health
Boston, Massachusetts

Mohd Famey Yussoff
Department of Environment, Ministry of Natural Resources and Environment
Putrajaya, Malaysia

Section I

Air Pollution and Health: Six Asian Case Studies

1 Air Pollution and Its Health Effects in China

Shaowei Wu and Xinbiao Guo

CONTENTS

1.1 BACKGROUND

1.1.1 General Information about China

China is the world's third largest country by land area and the world's most populous country. China has been among the world's fastest growing economies since economic liberalization began in 1978. China's annual average gross domestic product (GDP) growth between 1990 and 2013 exceeded 10% (Figure 1.1) (National Bureau of Statistics of China, 2015). As of 2014, China is the world's second largest economic entity after the United States, with the nominal GDP totaling approximately USD 10.3554 trillion. Meanwhile, the urbanization process in China has been increasing

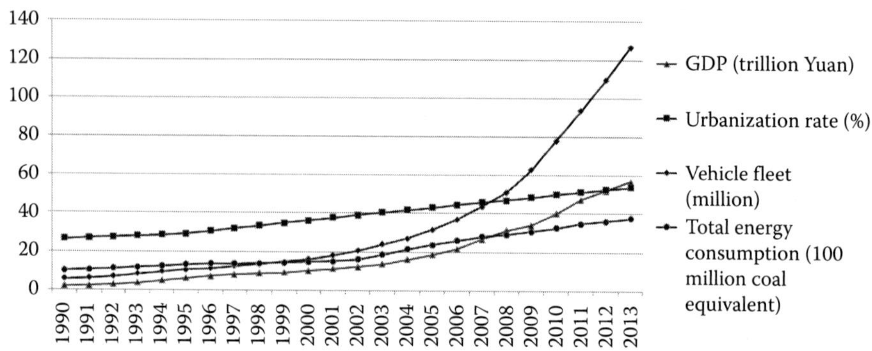

FIGURE 1.1 GDP, urbanization rate, total energy consumption, and vehicle fleet in China during 1990–2013.

in the past few decades. Only about 26% of the country's population lived in urban areas in 1990, whereas this percent has markedly increased to 53.7% in 2013. China now has more than 160 cities with a population of over one million, most of which are located in east and south China, and it is estimated that the urban population in China will reach one billion by 2030 (McKinsey Global Institute, 2009). Along with the rapidly expanding population in the urban areas, motorization becomes another epiphenomenon of the urbanization process. The scale of the civil vehicle fleet has reached 126.7 million by the end of 2013, which is more than five times higher than that of 10 years ago (National Bureau of Statistics of China, 2015).

1.1.2 Air Pollution Trend and Episodes in China

Along with the profound socioeconomic development, China has become the world's largest energy consumer in 2010 and accounts for 21% of global energy use (Badkar, 2012). China's total energy use was 3617.32 million tons of coal equivalents in 2012. Coal and oil, two major forms of fossil fuels, accounted for 66.6% and 18.8% of the total energy demands in China in 2012, respectively. As a result, combustion of fossil fuels has become the major source of urban ambient air pollution in China, and air pollution levels remain relatively high over recent years. For example, a recent source appointment study suggests that over 70% of ambient particulate matter with an aerodynamic diameter of 2.5 μm or less ($PM_{2.5}$) in Beijing (the capital of China) in 2010–2011 was related to the consumption of fossil fuels (Wu et al., 2014c). According to the governmental report, the annual average concentrations of several major air pollutants in 74 major cities in 2013 were high, which were 118 μg/m^3 for particulate matter with an aerodynamic diameter of 10 μm or less (PM_{10}), 40 μg/m^3 for sulfur dioxide (SO_2), and 44 μg/m^3 for nitrogen dioxide (NO_2) (Ministry of Environmental Protection of China, 2014). Annual average levels of PM_{10} and SO_2 showed decreasing trends, whereas annual average levels of NO_2 remained relatively constant during recent years (2003–2012) in China's 31 provincial capitals. However, levels of these air pollutants increased recently, with annual average levels in 2013 higher than those in 2012 (Figure 1.2) (National Bureau of Statistics of China, 2015). Levels

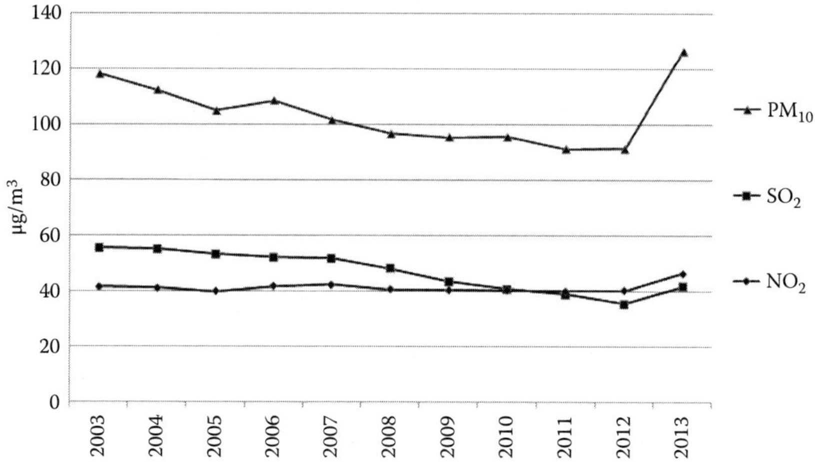

FIGURE 1.2 Average annual concentrations of three major air pollutants PM_{10}, SO_2, and NO_2 (µg/m^3) in 31 provincial capitals in China during 2003–2013.

of these air pollutants in China are still at the higher end of the world. Particularly, ambient levels of PM_{10} during 2003–2010 in 32 major Chinese cities ranked at the higher end (all rankings >800) among the world's 1,100 urban areas with a population above 100,000 inhabitants (World Health Organization, 2012); and eastern China was the world's hot spot with the highest $PM_{2.5}$ concentrations (values >100 µg/m^3 in its major industrial regions) during 2001–2006, based on satellite observations of the National Aeronautics and Space Administration of the United States (US NASA) (van Donkelaar et al., 2010).

Before 2012, China's national routine air-monitoring system did not include ambient $PM_{2.5}$, which is thought to be an air pollutant that is very dangerous to human health. Ambient $PM_{2.5}$ becomes a national standard air pollutant since the release of the modified national Ambient Air Quality Standards (AAQS) (Ministry of Environmental Protection of China, 2012). Among the 74 major cities with routine ambient $PM_{2.5}$ air-monitoring data in China in 2013, 69 cities (93%) did not reach the annual national AAQS II (35 µg/m^3), and none of the cities reached the annual national AAQS I (15 µg/m^3) (Figure 1.3).

Ambient air pollution levels have apparent seasonal patterns in China, with lower levels generally observed in warm season and higher levels generally observed in cold season. In the past few years, ambient air pollution in northern China became particularly severe in heating-season time (typically from late October to March), when the pollution emissions are enhanced and the weather conditions are not favorable for the diffusion of air pollutants. The levels of major air pollutants in urban areas often increase severalfold (in the magnitude of several hundreds of µg/m^3) of the average levels in normal years and can last for days, causing air pollution episodes of so-called haze or heavy smog. Sometimes the episodes are quite extensive, and other areas surrounding the heating areas may also be involved. The haze or smog during the episodes is often thick enough to block the sunlight from penetrating through the atmosphere and makes the sky looks gray in daytime. On the other

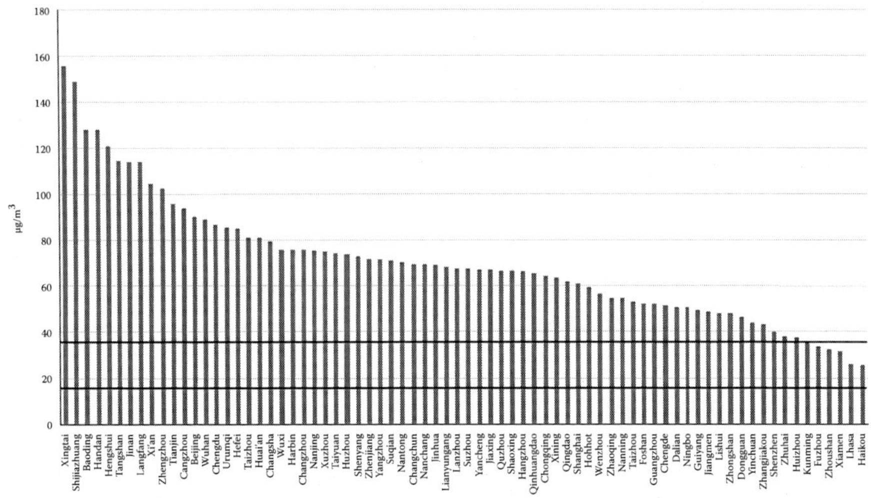

FIGURE 1.3 Annual average ambient $PM_{2.5}$ concentrations (µg/m³) in 74 major cities in China in 2013. The higher parallel line in the figure represents annual national Ambient Air Quality Standards (AAQS) II (35 µg/m³), and the lower parallel line represents annual national AAQS I (15 µg/m³).

hand, Asian dust storms from desert areas of China and Mongolia can also bring a large number of dry soil particles rich in heavy metals to eastern China during the spring season and elevate the ambient PM levels substantially. Exposure to ambient air pollution may cause significant health and economic burdens, which have been partly documented in a number of studies in recent years.

In summary, ambient air pollution has become a major environmental issue in China and has received great attention nationwide and worldwide. In this chapter, we review the current research literature on health effects, exposure measurements, risk assessment and management relevant to ambient air pollution in China, summarize the lessons learned, and recommend directions for further studies.

1.2 HEALTH EFFECTS

Ambient air pollution has been recognized as a major risk factor for population health in China. In a comprehensive study that performed analyses of disease burden for 231 diseases and injuries and 67 risk factors or clusters of risk factors relevant to China, ambient air pollution was estimated to cause 25,227 thousands of disability-adjusted life years in 2010, ranking fourth after dietary risk factors, high blood pressure (BP), and tobacco exposure (Yang et al., 2013). Another study suggests that an arbitrary policy, which provides free winter heating via the provision of coal for boilers in cities north of the Huai River but denies heat to the south, increases total suspended particles (TSP) air pollution and is causing the 500 million residents of northern China to lose more than 2.5 billion life-years of life expectancy (Chen et al., 2013d). The number of epidemiologic and toxicological studies conducted to investigate the health impact of ambient air pollution in China has been increasing during the most recent decade. Most of these studies focus on the short-term health effects within a few hours, days, or weeks after air pollution exposure, whereas potential long-term health effects over years have received less attention. Common study designs and health observational endpoints used in these studies include time-series and case-crossover studies examining the short-term effects of air pollution on clinically significant health outcomes (e.g., mortality and morbidity); cohort, cross-sectional, and ecological studies attempting to link long-term air pollution exposure with adverse health outcomes; panel, experimental, and cross-sectional studies investigating the subclinical effects of air pollution (e.g., changes in health indicators such as lung function and BP); and toxicological studies examining the biological mechanisms of air pollution action.

1.2.1 Ambient Air Pollution and Mortality and Morbidity

Disease mortality is the most extensively studied health endpoint in air pollution health studies in China. Disease morbidity measures, including hospital admissions, outpatient visits, and emergency room visits (ERVs), are less frequently investigated. Most of these epidemiologic studies have been focused on the short-term effects of air pollutants on mortality and morbidity within a few days, whereas only a few studies have been conducted to investigate the potential long-term effects of air pollutants on these health outcomes. In addition, other measures of disease burden, such as years of life lost and disability-adjusted life years, have also been used to assess the adverse health impact of air pollution (Guo et al., 2013; Zhang et al., 2006).

1.2.1.1 Short-Term Effects of Ambient Air Pollution on Mortality

Early epidemiologic studies investigating the association between day-to-day air pollution levels and mortality in China are generally restricted to one city or area. Meta-analyses and multicity studies comprising data from multiple Chinese cities have been increasing recently (Aunan and Pan, 2004; Chen et al., 2011a,b, 2012a,b,c; Lai et al., 2013; Shang et al., 2013; Wong et al., 2008; Tao et al., 2012; Tao et al., 2011; Lu et al., 2015). Locations most frequently studied are densely populated megacities, such as Beijing, Shanghai, and Hong Kong, where both routine air-monitoring system and mortality reporting system have been well established. There are also less frequent reports from other major Chinese cities including Guangzhou, Shenzhen, Wuhan, Chongqing, Shenyang, Xi'an, Taiyuan, Tianjin, and so on (Lai et al., 2013; Shang et al., 2013; Lu et al., 2015). PM_{10}, SO_2, and NO_2 are the three major air pollutants most commonly studied, whereas data on the other three major air pollutants $PM_{2.5}$, O_3, and carbon monoxide (CO) are still limited. Three general health outcomes, including total mortality, cardiovascular mortality, and respiratory mortality, are typical observational endpoints investigated in these epidemiologic studies. Table 1.1 summarizes the pooled risk estimates for daily mortality per unit increase of major air pollutants reported by meta-analyses and multicity studies in China.

In a multicity time-series study comprising data from 16 Chinese cities with population sizes ranging from 1.2 million to 12.3 million, city-specific average PM_{10} levels were reported to be 52–156 μg/m^3 during the study periods varying between 1996 and 2008, and a 10 μg/m^3 increase in 2-day moving-average PM_{10} was associated

TABLE 1.1
Summary of the Pooled Estimated Percent Increase in Risk of Daily Mortality Associated with Major Air Pollutants Reported in Meta-Analyses and Multicity Studies in China

	Primary Reference	Exposure, Increment	Percent Increase in Mortality (95% Confidence Interval/Posterior Interval)		
			Total	Cardiovascular	Respiratory
Meta-analysis of 6 studies[a]	(Aunan and Pan, 2004)	PM_{10}, 1 μg/m^3	0.03 (0.01, 0.05)	0.04 (0.02, 0.06)	0.06 (0.02, 0.10)
		SO_2, 1 μg/m^3	0.04 (0.02, 0.06)	0.04 (0.02, 0.06)	0.04 (0.02, 0.06)
Meta-analysis of 33 studies[a]	(Shang et al., 2013)	PM_{10}, 10 μg/m^3	0.32 (0.28, 0.35)	0.43 (0.37, 0.49)	0.32 (0.23, 0.40)
		$PM_{2.5}$, 10 μg/m^3	0.38 (0.31, 0.45)	0.44 (0.33, 0.54)	0.51 (0.30, 0.73)
		CO, 1 mg/m^3	3.70 (2.88, 4.51)	4.77 (3.53, 6.00)	-
		SO_2, 10 μg/m^3	0.81 (0.71, 0.91)	0.85 (0.70, 1.00)	1.18 (0.83, 1.52)
		NO_2, 10 μg/m^3	1.30 (1.19, 1.41)	1.46 (1.27, 1.64)	1.62 (1.32, 1.92)
		O_3, 10 μg/m^3	0.48 (0.38, 0.58)	0.45 (0.29, 0.60)	0.73 (0.49, 0.97)
Meta-analysis of 26 studies[a]	(Lai et al., 2013)	PM_{10}, 10 μg/m^3	0.31 (0.22, 0.41)	0.49 (0.34, 0.63)	0.57 (0.40, 0.75)
		NO_2, 10 μg/m^3	1.40 (1.06, 1.74)	1.62 (1.18, 2.05)	2.20 (1.56, 2.84)
		SO_2, 10 μg/m^3	0.71 (0.45, 0.97)	0.72 (0.39, 1.05)	1.29 (0.58, 1.99)
		O_3, 10 μg/m^3	0.42 (0.31, 0.53)	0.51 (0.25, 0.77)	0.48 (0.19, 0.76)
Meta-analysis of 36 studies[a]	(Lu et al., 2015)	PM_{10}, 10 μg/m^3	0.36 (0.26, 0.46)	0.36 (0.24, 0.49)	0.42 (0.28, 0.55)
		$PM_{2.5}$, 10 μg/m^3	0.40 (0.22, 0.59)	0.63 (0.35, 0.91)	0.75 (0.39, 1.11)
3 cities	(Wong et al., 2008)	PM_{10}, 10 μg/m^3	0.37 (0.21, 0.54)	0.44 (0.19, 0.68)	0.60 (0.16, 1.04)
		NO_2, 10 μg/m^3	1.19 (0.71, 1.66)	1.32 (0.79, 1.86)	1.63 (0.62, 2.64)
		SO_2, 10 μg/m^3	0.98 (0.74, 1.23)	1.09 (0.72, 1.47)	1.46 (0.84, 2.08)
		O_3, 10 μg/m^3	0.31 (0.13, 0.48)	0.29 (−0.09, 0.68)	0.23 (−0.22, 0.68)

(Continued)

TABLE 1.1 (*Continued*)
Summary of the Pooled Estimated Percent Increase in Risk of Daily Mortality Associated with Major Air Pollutants Reported in Meta-Analyses and Multicity Studies in China

	Primary Reference	Exposure, Increment	Percent Increase in Mortality (95% Confidence Interval/Posterior Interval)		
			Total	Cardiovascular	Respiratory
3 cities	(Chen et al., 2011b)	CO, 1 mg/m^3	2.89 (1.68, 4.11)	4.17 (2.66, 5.68)	-
4 cities	(Tao et al., 2011)	CO, 0.5 ppm	3.04 (2.18, 3.90)	3.62 (2.20, 5.06)	3.72 (1.71, 5.76)
4 cities	(Tao et al., 2012)	PM_{10}, 10 μg/m^3	0.79 (0.62, 0.96)	0.91 (0.64, 1.19)	1.26 (0.88, 1.65)
		NO_2, 10 μg/m^3	1.95 (1.62, 2.29)	2.12 (1.58, 2.65)	3.48 (2.73, 4.23)
		O_3, 10 μg/m^3	0.81 (0.63, 1.00)	1.01 (0.71, 1.32)	1.33 (0.89, 1.76)
16 cities	(Chen et al., 2012b)	PM_{10}, 10 μg/m^3	0.35 (0.18, 0.52)	0.44 (0.23, 0.64)	0.56 (0.31, 0.81)
17 cities	(Chen et al., 2012c)	NO_2, 10 μg/m^3	1.63 (1.09, 2.17)	1.80 (1.00, 2.59)	2.52 (1.44, 3.59)
17 cities	(Chen et al., 2012a)	SO_2, 10 μg/m^3	0.75 (0.47, 1.02)	0.83 (0.47, 1.19)	1.25 (0.78, 1.73)
32 cities	(Zhou et al., 2015b)	PM_{10}, 10 μg/m^3	-	-	1.05 (0.08, 2.04)

[a] The numbers of studies used for pooled estimates varied for different air pollutants and outcomes examined in the analysis.

with a 0.35% (95% posterior interval [PI]: 0.18, 0.52) increase of total mortality, a 0.44% (95% PI: 0.23, 0.64) increase of cardiovascular mortality, and a 0.56% (95% PI: 0.31, 0.81) increase of respiratory mortality in the combined analysis (Chen et al., 2012b). Two additional time-series analyses using data of 17 cities from the same study group reported city-specific average NO_2 levels between 23 and 67 μg/m^3 and SO_2 levels between 18 and 100 μg/m^3 during the study periods, and a 10 μg/m^3 increase in 2-day moving-average NO_2 was found to be associated with a 1.63% (95% PI: 1.09, 2.17), 1.80% (95% PI: 1.00, 2.59), and 2.52% (95% PI: 1.44, 3.59) increase of total, cardiovascular, and respiratory mortality, respectively (Chen et al., 2012c); whereas a 10 μg/m^3 increase in 2-day moving-average SO_2 was found to be associated with a 0.75% (95% PI: 0.47, 1.02), 0.83% (95% PI: 0.47, 1.19), and 1.25% (95% PI: 0.78, 1.73) increase of total, cardiovascular, and respiratory mortality, respectively (Chen et al., 2012a).

Most of the available studies are only able to examine the cardiovascular and respiratory mortality as two general health endpoints. Studies focusing on mortality due to specific cardiovascular or respiratory diseases and other cause-specific mortality are limited. Specific diseases that have been investigated in the epidemiologic studies include heart disease (Kan et al., 2010; Breitner et al., 2011; Tao et al., 2012; Wong et al., 2002b; Xu et al., 2014); cerebrovascular disease, including stroke (Chen et al., 2013a; Tao et al., 2012; Kan et al., 2010; Wong et al., 2002b); diabetes (Kan et al., 2004); acute respiratory infections, including pneumonia and influenza (Wong et al., 2002b, 2010; Kan et al., 2010); and chronic obstructive pulmonary disease (COPD) (Tao et al., 2012; Kan et al., 2010; Wong et al., 2002b). There are also sporadic reports on other cause-specific mortality due to cancer (Venners et al., 2003) and accident (Kan et al., 2010).

The levels of major air pollutants (especially ambient PM) reported in studies from China are generally higher than those reported in studies from the developed countries in North America and Europe. For example, a major study using data from 108 counties in the United States reported that the median of PM_{10} levels over the study areas during 1999–2005 was as low as 23.5 μg/m^3 (interquartile range [IQR]: 20.6, 28.6) (Peng et al., 2008). In contrast, the changes in daily mortality associated with per-unit increase in air pollution levels tend to be lower in China than those reported in developed countries (often >0.50% per 10 μg/m^3 increase of PM_{10} for total mortality) (Pope and Dockery, 2006; Samet et al., 2000). This difference may be explained by different characteristics of the study contexts, such as heterogeneity in local air pollution levels and sources, chemical constituents and associated toxicity, population sensitivity, age distribution, and socioeconomic factors (Kan et al., 2012; Health Effects Institute [HEI], 2010). Particularly, evidence from developed countries suggests that the exposure–response relationships for air pollution and daily mortality often tend to become flat at high concentrations (Pope et al., 2011). This flattening out of mortality risk at high exposure levels has been implicated in some studies (Chen et al., 2012c; Kan et al., 2007), though other studies also suggest a continuous increasing trend in mortality risk at higher exposure levels in China (Cao et al., 2012; Chen et al., 2013a).

Among the major air pollutants, ambient PM has received increasing attention because of greater potentials to induce adverse health effects. However, the uncertainty of the PM's health impact is substantial because of their varying sizes and chemical constituents, both of which may affect related biological toxicity critically (Valavanidis et al., 2008). Epidemiologic studies investigating the associations between daily mortality and size-fractioned PM and its chemical constituents are still rare in China. An earlier time-series study in Shanghai demonstrated that $PM_{2.5}$ was significantly associated with increased mortality, whereas coarse particles ($PM_{10-2.5}$) were not (Kan et al., 2007). Two later time-series studies further demonstrated that particles with smaller sizes were more strongly associated with daily mortality than were larger particles (Meng et al., 2013; Breitner et al., 2011). For example, one study found that IQR increases in particle-number concentrations of 0.25–0.28 μm, 0.35–0.40 μm, and 0.45–0.50 μm particles were associated with 2.41% (95% confidence interval [CI]: 1.23, 3.58%), 1.31% (95% CI: 0.52, 2.09%), and 0.45% (95% CI: 0.04, 0.87%) higher total mortality, respectively, in a Chinese city Shenyang (Meng et al., 2013). In contrast, there was little association between larger particles (with sizes of 1.0–2.5 μm and 2.5–10 μm) and total mortality. Data from a severely polluted Chinese city, Xi'an, suggest that several secondary and combustion-related chemical constituents of $PM_{2.5}$, including ammonium, nitrate, elemental carbon (EC), chlorine, and nickel, appeared to be most strongly associated with daily mortality (Cao et al., 2012; Huang et al., 2012b).

A few studies also reported the modification of the air pollution–daily mortality association by other covariates, such as demographic and socioeconomic characteristics of the study subjects and seasonal and meteorological factors (e.g., ambient temperature). Generally, air pollution–mortality association appears to be stronger among elders with age above 60, and subjects with low educational attainment (illiterate and primary school) (Kan et al., 2008; Chen et al., 2012b); appears to be stronger in warm season than in cold season (Kan et al., 2008; Qian et al., 2010); may vary by study locations (Chen et al., 2012b; Zhou et al., 2015b); and may be modified by changes in ambient temperature (Li et al., 2011).

1.2.1.2 Short-Term Effects of Ambient Air Pollution on Morbidity

Epidemiologic studies assessing the short-term effects of air pollutants on disease morbidity in China have been increasing in recent years. Most of these studies have been limited to a single city (e.g., Hong Kong, Beijing, or Shanghai). Compared with disease mortality, the reporting system for disease morbidity has not been well established in most Chinese cities. Hospital admissions, outpatient visits (clinic consultations), and ERVs are the most common morbidity measures examined in the epidemiologic studies. A few studies also investigated potential adverse reproductive outcomes associated with medium-term (several weeks to a few months) air pollution exposure.

Among the three cities (Hong Kong, Beijing, and Shanghai) which are most frequently investigated for the air pollution–morbidity association, Hong Kong provides a majority of such data available to date, probably owing to its well established health care system and enhanced governmental supervision (Lai et al., 2013; Wong et al., 2010). Morbidity due to respiratory diseases is the most commonly examined health endpoint in Hong Kong studies. Specific diseases investigated include respiratory tract infections (Tam et al., 2012b, 2014; Tian et al., 2013; Wong et al., 1999, 2010), asthma (Ko et al., 2007b; Lee et al., 2006; Wong et al., 1999, 2002a), COPD (Ko et al., 2007a; Tam et al., 2012b; Wong et al., 1999), ischemic heart disease (Qiu et al., 2013; Tam et al., 2012a; Wong et al., 1999, 2002a), heart failure (Tam et al., 2012a; Wong et al., 1999), cerebrovascular disease (Wong et al., 1999), and so on. Generally, children of ages <15 and elders of age ≥ 65 are more susceptible subjects than adults aged 15–65 (Wong et al., 2010; Ko et al., 2007b). A few studies also reported the short-term effects of Asian dust storms on emergency hospital admissions due to respiratory and cardiovascular diseases in Hong Kong (Tam et al., 2012a,b).

Data on the association between ambient air pollution and disease morbidity in mainland China are mainly contributed by studies from megacities such as Beijing and Shanghai. For example, several time-series and case-crossover studies in Beijing reported positive associations between major air pollutants and daily outpatient visits and ERVs based on records from a major hospital (Guo et al., 2009, 2010a,b; Xu et al., 1995); a cohort study reported a positive association between maternal exposures to SO_2 and TSP during the third trimester of pregnancy and low infant birth weight ($<$2,500 g) among 74,671 first-parity live births in Beijing (Wang et al., 1997); several time-series studies in Shanghai reported positive associations between major air pollutants and daily hospital admissions, outpatient visits, and ERVs (Cao et al., 2009; Chen et al., 2010; Hua et al., 2014; Wang et al., 2013; Zhao et al., 2014a); and another time-series study reported a positive association between 8-week averaged air pollution levels and preterm birth in Shanghai (Jiang et al., 2007). There are also reports from other Chinese cities such as Guangzhou, Shenzhen, and Lanzhou (Peng et al., 2011; Tao et al., 2014b; Yang et al., 2014).

According to a meta-analysis on the association between ambient air pollution and daily morbidity, the pooled relative risks ranged from 1.0021 (cardiovascular, PM_{10}) to 1.0162 (asthma, O_3) for five cause-specific hospital admissions (Lai et al., 2013). Data for outpatient visits and ERVs in China are not sufficient for a meaningful meta-analysis. Notably, the associations between air pollutants and daily morbidity rates are less consistent when compared with those between air pollutants and daily mortality rates. Although most epidemiologic studies reported positive associations between air pollutants and daily morbidity rates, significant inverse associations have also been reported for air pollutants and hospital admissions (Lai et al., 2013; Tian et al., 2013).

Only a few epidemiologic studies have investigated the associations of certain PM chemical constituents and size-fractioned PM with morbidity in China. Two time-series studies in Shanghai reported the associations of daily hospital visits and admissions with black carbon (BC), a combustion-related constituent (Wang et al., 2013; Hua et al., 2014). Both studies showed that BC appeared to be more strongly associated with health outcomes as compared with PM variables. Another time-series study examined the associations of 10 $PM_{2.5}$ chemical constituents and ERVs in Shanghai and found consistent positive associations of ERVs with organic carbon (OC) and EC, both of which are from the combustion of fossil fuel (Qiao et al., 2014). A few analyses also reported the association between various size-fractioned PM and ERVs due to respiratory and cardiovascular diseases in China (Leitte et al., 2011; Liu et al., 2013a). Particles with smaller sizes were found to be more strongly associated with daily ERVs due to cardiovascular disease in a Beijing hospital (Liu et al., 2013a). In contrast, smaller particles were not found to be more strongly associated with daily respiratory ERVs (Leitte et al., 2011) as well as daily mortality (Meng et al., 2013) than larger particles. Whether these findings suggest heterogeneous effects of smaller particles on cardiovascular and respiratory health require further investigation.

Beijing held the 29th Olympic Games in 2008 summer, and the government implemented a series of air pollution control measures during the Olympics through which the air pollution levels were reduced substantially (Wang et al., 2009c). One study based on a major hospital in Beijing observed a significant lower number of daily outpatient asthma visits during the Olympics than that before the Olympics (7.3 vs. 12.5 visits; relative risk = 0.54, 95% CI: 0.39, 0.75), suggesting a beneficial health effect related to air pollution reduction (Li et al., 2010).

Severe air pollution episodes, so-called haze or heavy smog, in the cold season have become a major public concern in China in recent years. The involved areas may encompass the entire central and eastern China during peak times of such episodes. Levels of major air pollutants (i.e., PM, SO_2, and NO_2) may all be elevated substantially during the episodes, and $PM_{2.5}$ has received special public attention because it is thought to be the major contributor of the haze events. Acute adverse health consequences, including increased hospital admissions, outpatient visits, and

ERVs following the episodes, have been frequently noted in the public press. The potential health impact of such heavy smog has been implicated in only a few scientific reports (Chen et al., 2013b; Zhou et al., 2015a). Further detailed assessments on the health burden associated with the air pollution episodes in the near future may help the government to tailor the relevant policies.

1.2.1.3 Long-Term Effects of Ambient Air Pollution

Generally, data from studies on the short-term effects of ambient air pollution provide important evidence for the daily air quality guidelines, whereas data from long-term follow-up studies comprise the main body of the evidence for the long-term (i.e., annual) air quality guidelines (US Environmental Protection Agency [USEPA], 2012; World Health Organization, 2006). Most studies investigating the long-term effects of ambient air pollution have been conducted in the developed countries in North America and Europe, where the air pollution levels are typically low and the pollutant ranges are relatively narrow (Dockery et al., 1993; Pope and Dockery, 2006; HEI, 2010). Only a handful of epidemiologic studies have assessed the long-term effects of ambient air pollution in China. Study designs used in these epidemiologic studies include cohort, cross-sectional, and ecological studies. Overall, the evidence for long-term effects of air pollution is still very limited in China.

Only several cohort studies have investigated the association between long-term air pollution and mortality in China. The first one used data from the China National Hypertension Survey, an established prospective cohort with 70,947 middle-aged adult participants in 16 provinces in China during 1991–2000 (Cao et al., 2011). The second study examined the association between long-term air pollution and mortality among 9,941 residents in Shenyang during 1998–2009 (Dong et al., 2012; Zhang et al., 2011b). The third study assessed the association of cardiovascular mortality with long-term exposure to PM_{10} among 39,054 subjects in four cities (including Shenyang) in northern China (Zhang et al., 2014). Another study investigated the association between long-term PM exposure and mortality among 71,431 middle-aged men during 1990–2006 (Zhou et al., 2014a). In these studies, participants' exposures to air pollutants were estimated according to fixed-site air-monitoring data averaged at the regional or community level, and significant positive associations were found between several major air pollutants and mortality due to cardiovascular and respiratory diseases. Notably, the magnitude of the associations varied substantially between studies. For example, a 10 $\mu g/m^3$ increase in TSP, SO_2, and nitrogen oxides (NO_X) were associated with a respective 0.9% (95% CI: 0.3, 0.5), 3.2% (95% CI: 2.3, 4.0), and 2.3% (95% CI: 0.6, 4.1) increase in cardiovascular mortality in the China National Hypertension Survey (Cao et al., 2011); whereas a 10 $\mu g/m^3$ increase in annual average concentrations of PM_{10}, SO_2, and NO_2 were found to be associated a respective 55% (95% CI: 1.51, 1.60), –4% (95% CI: –8, 1), and 146% (95% CI: 131, 163) increase in cardiovascular mortality among the Shenyang residents (Zhang et al., 2011b). The heterogeneous risk estimates may be associated with the different study contexts, as well as the unperfected study methodologies which may lead to exposure misclassification due to the use of regional- or community-level fixed-site air-monitoring data and information bias due to the retrospective characteristics of the data collection strategy.

Several cross-sectional studies and ecological studies also provide evidence for the associations or correlations between long-term air pollution levels and disease mortality and morbidity. One early ecological study found significant correlations between long-term airborne sulfate levels and total mortality and mortality due to cardiovascular disease, malignant tumor, and lung cancer in Beijing (Zhang et al., 2000); another ecological study using 52-year historical haze data between 1954 and 2006 found increases in lung cancer incidence and mortality following the air pollution events in Guangzhou (Tie et al., 2009); one recent ecological study also found significant correlations between annual NO_2 concentrations and lung cancer incidence and mortality in 10 Chinese cities (Huang et al., 2014b); one early cross-sectional study reported positive associations between long-term ambient PM levels and respiratory morbidity prevalence among 7,621 children in four Chinese cities (Zhang et al., 2002); and another cross-sectional study found that residential long-term exposure to air pollution was associated with increased risk of hypertension among 24,845 Chinese adults in three northeastern cities (Dong et al., 2013). However, the evidence from cross-sectional and ecological studies is thought to be inferior as compared with that from prospective cohort studies, given that the timing relationship between air pollution exposure and occurrence of the health endpoint is often unclear in cross-sectional and ecological studies. In addition, air pollution exposure observed in these studies are often averaged at the population level rather than at the individual level, thereby resulting in great uncertainty in exposure estimation and substantial bias in estimated risk. Nevertheless, such preliminary studies may provide clues for further more informative investigations.

1.2.2 Ambient Air Pollution and Subclinical Health Effects

Exposure to ambient air pollution may also cause a range of subclinical health effects in addition to the clinically significant health outcomes (i.e., mortality and morbidity). These effects can be reflected by changes in various health indicators, such as lung function, respiratory symptoms, heart rate variability (HRV), BP, and biomarkers measured in biological samples (e.g., blood, urine, exhaled breath condensate). A popular study design used in the investigation of the air pollution's subclinical effects is panel study. Panel studies usually include a few, several dozens of, or more than one hundred participants, who are observed repeatedly over time, thereby facilitating assessment of the health effects of changes in exposure over time (Janes et al., 2008). Methodologies including data collection and statistical analytic approaches used in panel studies are different from those used in traditional epidemiologic studies (e.g., time-series studies or cross-sectional studies). Furthermore, a few studies also used experimentally designed study protocols. In these studies, participants are intently exposed to different air pollution scenarios and measured for the same health variables under these scenarios, and the observed changes in health variables are then tested for difference statistically over the exposure scenarios. Nevertheless, the repeated assessment strategy of the panel and experimental study designs limits the number of participants that can be included in the investigation, and other study designs, including cross-sectional and cohort study designs, are used in larger-scale investigations. Studies on the subclinical effects of air pollution can provide important implications for the potential health impact of hazardous air pollutants and can serve as the bridge to link population-based epidemiologic studies documenting the adverse health effects of air pollution and toxicological studies documenting the biological mechanisms of air pollution action. Generally, several well-conducted panel and experimentally designed studies during the past few years have provided useful clues for the mechanistic pathways through which air pollution may promote the development of adverse respiratory and cardiovascular health outcomes, while traditional cross-sectional studies continue to contribute to the investigation of air pollution health effects in China.

1.2.2.1 Respiratory Health Effects of Air Pollution

The respiratory system is the first barrier that air pollutants encounter in the human body before entering the circulatory system and other parts of the body. Lung function is the most commonly investigated respiratory health indicator in association with air pollution in China, and most of the available studies are cross-sectional investigations among children (Liu and Zhang, 2009). Forced vital capacity (FVC), forced expiratory volume in 1 second (FEV_1), and maximal mid-expiratory flow rate (MMEF)/forced expiratory flow between the 25th and 75th percentile of forced vital capacity (FEF_{25-75}) are the three major lung function measures examined in these studies. A previous literature review based on data from 11 cross-sectional studies among 7,588 children in China reported significant aggregated reductions of –17.6 mL in FVC and –16.2 mL in FEV_1 per 10 μg/m^3 increase in TSP (Liu and Zhang, 2009). A few studies assessed the association between long-term air pollution exposure and lung function in children (Gao et al., 2013; He et al., 2010b). One study followed 1,983 children prospectively for 6 months in three districts with different air pollution levels in Guangzhou and performed twice lung function tests among the children (He et al., 2010b). Compared with children living in the least polluted district, children living in the most polluted district generally had significantly lower annual growth rates in lung function, as measured by FEV_1, forced expiratory flows at 25% (FEF_{25}), and FEF_{25-75}. Another cross-sectional study in Hong Kong provided risk estimates for lung function changes in 3,168 children associated with long-term PM_{10}, which was measured at the population level (Gao et al., 2013). In addition, one cross-sectional study reported significant inverse associations between short-term levels of smaller particles (PM_{10} and $PM_{2.5}$) and eight lung-function measures among 260 children in Beijing (Wang et al., 2010).

A panel study investigated the association between a number of air pollutants and 32 chemical constituents of $PM_{2.5}$ and lung function in a group of healthy volunteers (40 university students) in Beijing during 2010–2011, and provided the first hand information for the potential respiratory effects of various PM chemical constituents in the context of suburban and urban air pollution in China (Wu et al., 2013a,d, 2014b,c). Study participants were measured for lung function biweekly 12 times during three 2-month-long study periods before and after relocating from a suburban area to an urban area (one period in the suburban area and two periods in the urban area) with changing ambient air pollution levels and contents in Beijing. A subgroup of the participants also provided daily morning and evening peak expiratory flow (PEF) and FEV_1 measurements over 6 months. Study results suggest that $PM_{2.5}$ showed the most robust association with lung function among the major air pollutants measured in the study (e.g., PM_{10}, $PM_{2.5}$, $PM_{10-2.5}$, CO, NO_X, and NO_2), and several $PM_{2.5}$ metallic constituents (e.g., copper, cadmium, arsenic, stannum, calcium, and magnesium) showed consistent inverse associations with lung function measures (Wu et al., 2013a,d). Further analyses suggest that $PM_{2.5}$ from dust/soil and industry sources, which were enriched by these metallic constituents, were most strongly associated with the reductions in lung function measures (Wu et al., 2014c). Interestingly, the inverse associations between $PM_{2.5}$ and lung function measures appeared to be stronger at longer exposure metrics up to 1–2 weeks as compared with those at shorter exposure metrics within a few days, suggesting a potential cumulative exposure effect over time (Wu et al., 2013d, 2014b). In addition, the air pollution–lung function association appeared to be modified by ambient temperature, suggesting that it is necessary to account for temperature levels in the assessment of short-term respiratory effects of air pollution exposure (Wu et al., 2014b). Another repeated-measure study among 60 truck drivers and 60 office workers measured their postwork lung function and personal $PM_{2.5}$ exposures on two separate days, and found significant inverse associations of FEV_1 and FVC with several crustal metals (i.e., silicon, aluminum, and calcium) among nine measured $PM_{2.5}$ elemental constituents (Baccarelli et al., 2014). Another panel study based on 40 days of continuously measured data on ambient PM and eight metallic constituents also found significant inverse associations between several metallic constituents (i.e, lead, nickel, iron, manganese, and chrome) and lung function among 107 children in Baotou, an industrial city in northern China (Madaniyazi et al., 2013). Most of the identified metallic constituents are transition metals, which are thought to be able to stimulate the production of reactive oxygen species when delivered to the airways, and then induce airway injury and inflammation followed by a series of cardiopulmonary responses (Gonzalez-Flecha, 2004). However, a recent randomized, double-blind crossover trial among 35 healthy college students in Shanghai did not observe apparent improvements in lung function after reducing indoor particles of outdoor origin using air purifiers for 48 hours (Chen et al. 2015a). The reason for nonsignificant change in lung function may be due to the relatively short intervention period.

Several cross-sectional studies also reported positive associations between long-term air pollution exposure and the occurrence of respiratory symptoms in children (Liu et al., 2013b; Pan et al., 2010; Zhang et al., 2002) or adults (Wang et al., 2011). The major respiratory symptoms found to be associated with air pollution include persistent cough, persistent phlegm, wheezing, and current asthma. However, air pollution levels in these cross-sectional studies were measured at the population level, and thus limit the reliability of the estimated respiratory effects of air pollution.

In addition, several studies also assessed the potential respiratory effects of air pollution using biomarkers determined in human biological samples (Chen et al., 2015a; Lin et al., 2011; Huang et al., 2012c). As aforementioned, stringent air pollution control reduced the air pollution levels during the 2008 Beijing Olympics substantially, providing a unique opportunity to observe the air pollution effects over time. One panel study examined the association between ambient $PM_{2.5}$ and BC levels and exhaled nitric oxide (eNO), an acute respiratory inflammation biomarker, among 36 children over five visits before and during the Olympics (Lin et al., 2011). The study found that both air pollution concentrations and eNO were clearly lower during the

Olympics, and 24-hour average BC and $PM_{2.5}$ concentrations showed significant positive associations with eNO. The association between BC and eNO also appeared to be stronger than that between $PM_{2.5}$ and eNO. Another panel study also investigated the relationship between air pollution levels and a group of biomarkers reflecting pulmonary inflammation and pulmonary and systemic oxidative stress (i.e., eNO, exhaled breath condensate markers, and urinary 8-hydroxy-2-deoxyguanosine) among 125 healthy adults over six visits in the pre-(high pollution), during-(low pollution), and post-Olympic (high pollution) periods (Huang et al., 2012c; Gong et al., 2014). Study results suggest significant decreases in the biomarker levels from the pre- to the during-Olympic period and significant increases in the same biomarker levels from the during-Olympic to the post-Olympic period. Statistical analyses also found consistent positive associations between increased pollutant concentrations and biomarker levels. A further analysis from the study showed similar changes for malondialdehyde (MDA), a biomarker of oxidative stress, in exhaled breath condensate samples of the study subjects (Gong et al., 2013). In addition, the intervention trial among Shanghai college students also found a significant decrease in eNO associated with the use of air purifiers (Chen et al., 2015a). These findings support the important roles of oxidative stress and pulmonary inflammation in mediating air pollution health effects.

1.2.2.2 Cardiovascular Health Effects of Air Pollution

The cardiovascular system is another human body system that is susceptible to air pollution, and cardiovascular health effects have been the hot spots in recent air pollution health studies in China. Several well-conducted panel studies have provided a general view for the potential subclinical cardiovascular effects that air pollution may induce. The most commonly investigated cardiovascular health indicator is HRV, followed by BP and circulating biomarkers.

Changes in HRV can reflect altered function in autonomic nervous system (ANS), which is considered as one of the pathophysiologic pathways through which air pollution influences the cardiovascular system (Pope et al., 2004). An earlier panel study which followed 11 taxi drivers around the Beijing Olympics provided the first detailed assessment on the subclinical health effects of air pollution changes around the Olympics (Wu et al., 2010). This study measured taxi drivers' in-car exposure to $PM_{2.5}$ and gaseous air pollutants and ambulatory electrocardiography (ECG) in a 12-hour work shift in each of the before, during, and after the Olympic periods, and found significant improvements in drivers' HRV levels during the Olympics, when the air pollution levels were lower. In contrast, the HRV levels were much lower when the air pollution levels were higher, before and after the Olympics. Among the air pollutants, $PM_{2.5}$ was found to have the strongest inverse association with changes in 5-min HRV indices during the measurement periods. Further analyses revealed significant associations between acute CO exposure and 5-min HRV (Wu et al., 2011b) and between certain $PM_{2.5}$ chemical constituents and 12-hour HRV (Wu et al., 2011a). Interestingly, the association between air pollution and HRV also appeared to be modified by temperature (Wu et al., 2013c). Significant improvements in HRV were also observed in another two panels of patients with cardiovascular disease during the Olympics, and ambient PM (PM_{10} or $PM_{2.5}$) levels were found to be significantly inversely associated with HRV changes among the patients (Huang et al., 2012d;

Jia et al., 2009; Xu et al., 2013). The association between air pollution and HRV levels may vary by subject characteristics such as body mass index, gender, systemic inflammation status, and disease status such as type 2 diabetes (Huang et al., 2012d; Sun et al., 2015). Another panel study also reported an inverse association between short-term O_3 exposure and HRV among 20 healthy elders in Beijing (Jia et al., 2011a). Notably, two panel studies among healthy elderly or young subjects found evidence for the positive associations between air pollution and HRV changes (Wu et al., 2010; Jia et al., 2012b), suggesting a potential heterogeneous pattern of cardiac response to external stimuli among healthy subjects as compared with that among susceptible subjects. Several randomized, crossover intervention studies in Beijing also observed higher HRV levels under exposure scenarios with lower air pollution levels (in parks or with a highly efficient facemask) and lower HRV levels under exposure scenarios with higher air pollution levels (in traffic centers or without the facemask) in either healthy subjects or patients with coronary heart disease (Huang et al., 2013; Langrish et al., 2009; Langrish et al., 2012). Specifically, one of the studies exposed 40 healthy participants to two different scenarios (traffic center and park) for 2 hours in two separate occasions and found amplified air pollution effects on HRV at high noise levels, (>65.6 A-weighted decibels (dB[A])) than at low noise levels, (<65.6 dB[A]), suggesting that noise is an important modifier of the air pollution effects on HRV (Huang et al., 2013). In addition, another panel study also found evidence for the association between size-fractioned ultrafine particles and BC and autonomic dysfunction in 53 subjects with diabetes or impaired glucose tolerance, and the association tended to be larger for particles with smaller size fractions (Sun et al., 2015).

BP is another cardiovascular health indicator frequently examined in air pollution health studies. The aforementioned study among truck drivers and office workers measured their postwork BP twice and found significant positive associations of ambient PM_{10} with systolic, diastolic, and mean BPs in the study subjects (Baccarelli et al., 2011). Two other repeated-measure studies reported significant increases in ambulatory BP levels in association with short-term $PM_{2.5}$ and BC exposures within minutes and hours among patients with cardiovascular disease or metabolic syndrome (Huang et al., 2012d; Zhao et al., 2014b). The aforementioned panel study among 125 healthy adults around the Olympics found a significant increase in systolic BP from during the Olympics to post-Olympics, whereas the association pattern with multiple pollutants was not consistent (Rich et al., 2012). A randomized crossover study found significant lower systolic BP in 15 healthy subjects during 2-hour city walk with a highly efficient facemask versus without the facemask (Langrish et al., 2009), and a later similarly designed study also found significant lower mean arterial pressure in 98 patients with coronary heart disease under the exposure scenario with the facemask versus without the facemask (Langrish et al., 2012). A panel study among 35 diabetes patients with six repeated measurements found that the association of systolic BP and pulse pressure with airborne PM strengthened with decreasing diameter, and these effects occurred immediately even after 0-2h and lasted for up to two days following exposure (Zhao et al., 2015). These findings provide evidence for potentially important size and temporal patterns of airborne PM in elevating BP among susceptible individuals. The aforementioned panel study among 40 university students who relocated between the Beijing suburban and urban areas provided a detailed investigation for the associations between various air pollutants and $PM_{2.5}$ chemical constituents and BP changes observed over the study (Wu et al., 2013b). Study results suggest that several combustion-related constituents (i.e., EC, OC, chloride) and metallic constituents (i.e., nickel, zinc, magnesium, lead, and arsenic) showed the most consistent positive associations with BP variables. Further analyses revealed that $PM_{2.5}$ from coal combustion, which was enriched by carbonaceous fractions and chloride, was most strongly associated with the increased BP in the study subjects (Wu et al., 2014c). In addition, this study also found significant interactions between major air pollutants and temperature, as evidenced by stronger estimated air pollution effects on BP at low temperature (<median) than at high temperature (≥median) (Wu et al., 2015b). Furthermore, the intervention trial among Shanghai college students found significant reductions in BP after using air purifiers (Chen et al., 2015a). In addition, there are also several reports on long-term exposure to air pollution and BP from large cross-sectional studies. One cross-sectional study reported a potential association between long-term air pollution and increased BP among 24,845 Chinese adults in three northeastern cities (Dong et al., 2013). A further cross-sectional study with 9,354 children found stronger associations between long-term air pollution exposure and BP and hypertension in obese/overweight children than those in normal weight children, suggesting that obesity may amplify the adverse effects of air pollution on BP and hypertension (Dong et al., 2015).

Several panel studies also collected repeated blood samples over time among the study subjects and attempted to quantify the association between air pollution exposure and levels of circulating biomarkers which could reflect cardiovascular health status. One study measured 24-hour personal $PM_{2.5}$ exposure and a group of inflammatory and immune biomarkers twice among 110 traffic policemen in Shanghai and found that higher $PM_{2.5}$ exposure was associated with increases in high-sensitivity C-reactive protein (CRP), IgG, IgM, and IgE, and decreases in IgA and CD8 T cells (Zhao et al., 2013). The panel study among 125 healthy adults examined several biomarkers of systemic inflammation (CRP, fibrinogen, and white blood cell count) and thrombosis or endothelial dysfunction (soluble P-selection, soluble CD40 ligand, and von Willebrand factor [vWF]) over six study visits and found generally similar changing trends in levels of these biomarkers and air pollutants around the Olympics (Rich et al., 2012). Statistical analyses further demonstrated significant associations between air pollution and biomarker levels, suggesting a potential effect of air pollution on cardiovascular physiology in healthy young persons (Rich et al., 2012; Gong et al., 2014). A further analysis from this study found varying levels of vWF according to vWF gene polymorphism, which may have affected the participants' response to air quality improvements (Yuan et al., 2013). The panel study among 40 university students in Beijing collected 12 repeated blood samples over the study and provided a detailed assessment for the associations between various air pollutants and $PM_{2.5}$ chemical constituents and biomarkers of inflammation, coagulation, and homocysteine (Wu et al., 2012). Several metallic constituents (i.e., zinc, stannum, magnesium, iron, titanium, cobalt) were found to have robust positive associations with levels of inflammatory biomarkers (tumor necrosis factor alpha [TNF-a] and fibrinogen), and further analyses revealed positive associations between levels of these biomarkers and $PM_{2.5}$ from secondary sulfate/nitrate and dust/soil, which were enriched by the identified metallic constituents (Wu et al., 2014c). In contrast, the associations of coagulation and homocysteine biomarkers with air pollutants have been less consistent Nevertheless, a recent panel study in 34 healthy young adults in Shanghai found significant increases in biomarkers of inflammation (e.g., CRP, fibrinogen, monocyte chemoattractant protein-1), coagulation (vWF, plasminogen activator inhibitor-1, CD40 ligand) and vasoconstriction (endothelin-1) in association with PM with smaller size within 24 hours of exposure (Chen et al. 2015b). The aformentioned intervention trial among college students also found significant decreases in several inflammation and coagulation biomarkers after using air purifiers, suggesting a consistent pattern of cardiovascular benefits associated with air quality improvement (Chen et al., 2015a).

1.2.2.3 Other Subclinical Health Effects of Air Pollution

A few studies also investigated the associations between air pollution and other subclinical health effects, such as alterations in neurobehavioral function, systemic oxidative stress, micronuclei frequency, and epigenetics. For example, a cross-sectional study performed multiple neurobehavioral function tests (e.g., line discrimination, visual retention, continuous performance, digit Symbol, pursuit Aiming, sign register) among 431 schoolchildren from a clean area with low air pollution levels and 430 schoolchildren from a polluted area with high air pollution levels in Quanzhou and found that children from the polluted area showed poor performance on all neurobehavioral function tests (Wang et al., 2009a); a study in Beijing repeatedly measured two security guards' exposures to ambient $PM_{2.5}$ and related chemical species at the work site near a heavy traffic road and their pre- and post-work-shift urinary levels of 8-hydroxy-2′-deoxyguanosine (8-OHdG, a biomarker of oxidative stress and DNA damage) for 29 days, and found that post-work-shift urinary 8-OHdG concentrations were significantly and positively associated with $PM_{2.5}$, polycyclic aromatic hydrocarbons (PAHs), and metals, whereas pre-work-shift 8-OHdG concentrations were only associated with $PM_{2.5}$ at the background site (Wei et al., 2009, 2010); a study among 110 policemen and 110 controls measured their 24-hour personal $PM_{2.5}$ exposures and blood benzo[a]pyrene (BaP) 7,8-diol-9,10-epoxide-DNA adducts (another biomarker of DNA damage) and urinary 1-hydroxypyrene (1-OHP, a metabolite of PAH exposure), and found that policemen had higher levels of these biomarkers compared with controls, and $PM_{2.5}$ exposure was associated with a 0.8% increase in blood BaP 7,8-diol-9,10-epoxide-DNA adducts and 1.1% increase in 1-OHP after adjusting for potential confounders (Li et al., 2014b); a panel study measured urinary MDA levels among 120 schoolchildren from two cities in China and two cities in Korea for five consecutive days, and found significant increases in urinary MDA levels associated with ambient PM levels (Bae et al., 2010); the panel study among 40 university students found that a subset of PM2.5 metals (e.g., iron and nickel) were more closely associated with increased plasma oxidized low-density lipoprotein, another biomarker of systemic oxidative stress, than major air pollutants (Wu et al., 2015c). the study among Beijing truck drivers and office workers also found a significant inverse association between ambient PM_{10} and shorter telomere length, a biomarker of cardiovascular risk that is modified by inflammation and oxidative stress (Hou et al., 2012), and an inverse association between air pollution and mitochondrial DNA copy number, a marker of mitochondrial damage and malfunctioning that is associated with various diseases or conditions (Hou et al., 2013); a cross-sectional study evaluated the genotoxic effects of ambient air pollution among 129 rural and industrial female residents in Shenyang using micronuclei assays and polymorphic analyses of metabolic enzyme and DNA repair genes, and found that industrial female residents who exposed to higher air pollution levels had a higher micronuclei frequency, a sensitive marker of DNA damage (Ishikawa et al., 2006); and several cross-sectional or longitudinal studies also reported a potential association between air pollution and adverse reproductive health, such as slower fetal and child growth and development and decreased semen quality in adult men (Tang et al., 2014; Zhou et al., 2014b).

1.2.3 Biological Mechanisms for the Health Effects of Ambient Air Pollution

While human subclinical health studies provide clues for the mechanistic pathways through which ambient air pollution may promote the development of adverse health outcomes (i.e., mortality and morbidity), more direct evidence for the biological mechanisms of air pollution action comes from toxicological studies. Toxicological studies include *in vitro* studies and *in vivo* studies, which expose cells and animals to air pollutants experimentally and compare the observed biological responses under different exposure conditions. The high doses used in the toxicological studies are usually able to induce a series of *in vitro* and *in vivo* biological responses in a dose-dependent manner, which may then help to clarify the mechanisms of air pollution action. To date, a majority of toxicological studies in China have been conducted to investigate the biological effects of ambient PM on the respiratory and cardiovascular systems. Among the ambient particles, $PM_{2.5}$ has been shown to possess stronger cytotoxicity than larger particles (Hsiao et al., 2000), and thus has received increasing attention in recent years. Table 1.2 summarizes the evidence for the biological pathways of air pollution action based on toxicological studies and subclinical health studies in China in recent years.

A significant body of evidence suggests that PM exposure may lead to increased pulmonary and systemic inflammation (Bai et al., 2007), and pulmonary inflammation may play a major role in enhancing the extrapulmonary translocation of particles (Chen et al., 2006). Exposure to ambient particles has been shown to increase the secretion of inflammatory factors (e.g., TNF-a, interleukins) in human and animal lung cells (Jia et al., 2006) and animal respiratory and circulatory systems (Chen et al., 2013c; Deng et al., 2010; Yoshida et al., 2010), cause lung and systemic inflammation by enhancing inflammatory cells recruitment (Liu et al., 2014a; He et al., 2010a), and induce airway injury in animals (Yoshida et al., 2010). Specifically, an *in vivo* study exposed mice to Beijing $PM_{2.5}$ and filtered air during and after the 2008 Beijing Olympics, and found that short-term increases in exposure to ambient $PM_{2.5}$ led to increased systemic inflammatory responses, as demonstrated by increased levels of several biomarkers (e.g., monocyte chemoattractant protein 1 and interleukin 6) and enhanced recruitment of macrophages and neutrophils in the lung, spleen, and visceral adipose tissue, whereas short-term air quality improvements were significantly associated with reduced overall inflammatory responses (Xu et al., 2012).

TABLE 1.2
Summary of Evidence for the Biological Pathways of Air Pollution Action Based on Toxicological Studies and Subclinical Health Studies in China

	Involved Body Systems	Evidence Strength[a] Short-Term (hours/days)	Evidence Strength[a] Longer-Term (months/years)	Related References
Inflammation	Respiratory/systemic	↑↑↑	↑	(Bao et al., 2013; Chen et al., 2006, 2013c, 2015a,b,c; Deng et al., 2010; Gong et al., 2014; He et al., 2010a; Huang et al., 2012c; Jia et al., 2006; Li et al., 2007; Lin et al., 2011; Liu et al., 2014a, 2014b; Rich et al., 2012; Wu et al., 2012, 2014c; Xu et al., 2012; Yoshida et al., 2010; Zhao et al., 2013; Ding et al., 2014)
Immune function	Systemic	↑	↑	(He et al., 2010a; Yoshida et al., 2010; Hong et al., 2013; Zhao et al., 2013; Liu et al., 2014a; Ding et al., 2014)
Oxidative stress	Respiratory/systemic	↑↑	↑	(Bae et al., 2010; Deng et al., 2013; Fan et al., 2012; Gong et al., 2013, 2014; Huang et al., 2012c; Liu et al., 2011, 2014b; Wei et al., 2009, 2010; Xu et al., 2011; Ding et al., 2014)
Blood pressure	Cardiovascular	↑↑↑	↑	(Baccarelli et al., 2011; Chen et al., 2015a Dong et al., 2013, 2015; Gong et al., 2014; Huang et al., 2012d; Langrish et al., 2009, 2012; Rich et al., 2012; Wu et al., 2013b, 2014c, 2015b; Zhao et al., 2014b, 2015)

(Continued)

TABLE 1.2 (*Continued*)
Summary of Evidence for the Biological Pathways of Air Pollution Action Based on Toxicological Studies and Subclinical Health Studies in China

	Involved Body Systems	Evidence Strength[a] Short-Term (hours/days)	Longer-Term (months/years)	Related References
Cardiac autonomic nervous system imbalance	Cardiovascular	↑↑↑	-	(Deng et al., 2009a; Huang et al., 2012d, 2013; Jia et al., 2012a, 2012b; Jia et al., 2009, 2011a; Langrish et al., 2009, 2012; Sun et al., 2015; Wu et al., 2010, 2011b, 2011a, 2013c; Xu et al., 2013)
Blood coagulation/ endothelial dysfunction	Cardiovascular	↑	-	(Chen et al., 2015a, b; Deng et al., 2010; Gong et al., 2014; Rich et al., 2012; Wu et al., 2012; Yuan et al., 2013)
Lipid metabolism/ atherosclerosis	Cardiovascular	-	↑	(Chen et al., 2013c; Liu et al., 2014a; Wu et al., 2012)
Nuclear factor-kB activation	Nervous	↑	-	(Sang et al., 2011)
DNA/ chromosome alteration	Genetic	↑	↑	(Hou et al., 2012, 2013; Hsiao et al., 2000; Ishikawa et al., 2006; Li et al., 2014b; Tang et al., 2014; Wei and Meng, 2006; Wei et al., 2009, 2010; Ding et al., 2014)
DNA methylation	Epigenetics	↑	-	(Byun et al., 2013; Chen et al., 2015c; Guo et al., 2014; Hou et al., 2014)
Cell cycle/ proliferation	Cytotoxic	↑	-	(Deng et al., 2007; Ding et al., 2014)
Cell communication	Cytotoxic	↑	-	(Wang et al., 2006)

[a] "-" indicates overall little evidence, "↑" indicates overall weak evidence, "↑↑" indicates overall moderate evidence, "↑↑↑" indicates overall strong evidence.

Exposure to other gaseous air pollutants, such as O_3 and SO_2, has also been found to cause increased lung inflammation by triggering expressions of inflammatory factors in animals (Bao et al., 2013; Li et al., 2007). Specifically, one study found that exposure to SO_2 affects the mRNA and protein expression levels of several asthma-related genes in asthmatic rats, suggesting that SO_2 pollution may aggravate asthma disease through increasing the expressions of relevant genes at the transcriptional and translational levels in the lungs and tracheas (Li et al., 2007, 2008).

In addition, involvement of the immune system (e.g., T-cell activation) has also been implicated in the animal's response to PM exposure (He et al., 2010a; Yoshida et al., 2010), and exposure of pregnant mice to PM can result in postnatal immune dysfunction by exacerbation of Thl/Th2 cell deviation in mice offspring (Hong et al., 2013). An *in vivo* study showed that $PM_{2.5}$ exposure can result in whole-body insulin resistance by regulating inflammation in visceral adipose tissues, hepatic lipid metabolism, and glucose utilization in skeletal muscle in mice via both CC-chemokine receptor 2 (CCR2, a molecule that plays a critical role in the entry of innate immune cells into tissue) –dependent and –independent pathways, providing new mechanistic links between air pollution and metabolic abnormalities (Liu et al., 2014a).

Oxidative stress is another pathway proposed to be involved in the air pollution effects. Exposure to PM_{10} has been shown to increase the MDA levels and decrease the activities of superoxide dismutase, an important antioxidant defense, in the serum, lung, and heart tissues in rats (Lu et al., 2011). Exposure to $PM_{2.5}$ can trigger oxidative stress in brown adipose tissues, as demonstrated by increased production of reactive oxygen species in brown adipose depots (Xu et al., 2011). $PM_{2.5}$-induced oxidative stress has been shown to trigger autophagy in human lung cells, which may contribute to the PM-induced impairment of lung function (Deng et al., 2013). An *in vitro* study also investigated the chemical constituents and pollution sources behind the $PM_{2.5}$-induced oxidative stress in human bronchial epithelial cells and found that a secondary source (e.g., sulfate and nitrate) explained the largest fraction of reactive oxygen species variability observed in experiments (Liu et al., 2014b).

Cardiac ANS function is a cardiovascular health indicator that is sensitive to external stimuli, as reflected by changes in HRV. An *in vivo* study found that intratracheal instillation could cause arrhythmia in rats (Deng et al., 2009a). Another *in vivo* study explored the cardiac effects of BC, a major constituent of ambient PM, and found significant decreases in several HRV indices at higher BC exposure doses in mice (Jia et al., 2012a). Interestingly, only slight pulmonary inflammation and myocardial injury were observed in exposed mice, suggesting that BC can disturb cardiac ANS function through mechanisms independent of apparent myocardial and pulmonary injury.

Exposure to ambient PM also affects coagulation function. Animals exposed to $PM_{2.5}$ showed decreased plasma prothrombin time and increased plasma levels of fibrinogen and tissue factor, suggesting a hypercoagulable status after exposure (Deng et al., 2010). Additionally, nickel sulfate–treated animals also showed similar changes in coagulation function, suggesting that nickel may play a role in PM-related coagulative effects.

Recent *in vivo* experiments have shown that $PM_{2.5}$ modulates lipid metabolism (Liu et al., 2014a), and results from another *in vivo* study suggest that PM exposure can accelerate atherosclerosis in animals (Chen et al., 2013c). The latter study investigated the long-term effects of PM exposure by exposing apolipoprotein E knockout mice to ambient PM and filtered air in Beijing for 2 months, and found that the serum total cholesterol and low-density lipoprotein levels of PM-exposed mice were significantly higher than those of filtered-air-exposed mice. Pathological analysis revealed that the plaques area in the aortic arch of the PM-exposed mice increased significantly as compared with that of the filtered-air-exposed mice, suggesting that PM exposure could contribute to the progression of atherosclerosis (Chen et al., 2013c).

The neurotoxicity of air pollution has also been investigated. A toxicological study showed that SO_2 exposure could produce a neuronal insult, and the neurotoxic effect was likely via stimulating cyclooxygenase-2 elevation by activation of nuclear factor (NF)-κB activity and its acting on the promoter-distal NF-κB-binding site of cyclooxygenase-2 promoter (Sang et al., 2011).

Several *in vitro* studies also investigated the cytotoxic effects of ambient PM, especially PM related to Asian dust storms (Deng et al., 2007; Wang et al., 2006; Wei and Meng, 2006). An early study found that solvent-extractable organic compounds of $PM_{2.5}$ could induce DNA damage (Hsiao et al., 2000), and a later study demonstrated that dust-storm $PM_{2.5}$ and its extract can induce increased chromosomal aberration (Wei and Meng, 2006). Other studies showed that both organic and inorganic extracts of dust storms and normal $PM_{2.5}$ could inhibit cell proliferation and induce cell arrest (Deng et al., 2007), whereas organic extract of $PM_{2.5}$ showed stronger inhibitory effects on the gap-junctional intercellular communication than inorganic extract (Wang et al., 2006).

An *in vitro* study demonstrated the first transcriptomic investigation in the air pollution research field in China using a genome-wide approach (Ding et al., 2014). The authors identified alterations in a series of genes and pathways after exposing human bronchial epithelial cells to different concentrations of $PM_{2.5}$ collected in a Chinese megacity. The altered genes included those involved in inflammatory and immune response, response to oxidative stress, and response to DNA damage stimulus. Pathway analysis revealed that different doses of $PM_{2.5}$ triggered partially common disturbed pathways for cellular processes, environmental information processing, genetic information processing, and metabolism. Flow cytometry assay suggested that there were statistically significant differences in the G1 phase of the cell cycle after low- or high-dose $PM_{2.5}$ exposure when compared with the unexposed controls, and only high-dose $PM_{2.5}$ significantly increased the proportion of cells in the S phase of the cell cycle. This study provides a novel approach to investigating the biological mechanisms underlying $PM_{2.5}$-induced adverse health effects.

Epigenetic changes associated with air pollution exposure have become a new research spot in recent years. Reduced levels of DNA methylation have been linked to aging, oxidative stress, and cardiovascular disease. Several additional analyses from the Beijing truck driver study provide the first effort to link air pollution and epigenetic changes in China. While small mitochondrial DNA methylation was not associated with traffic-related PM exposure among the study participants (Byun et al., 2013), significant associations between air pollution and methylation of three DNA tandem repeats (i.e., SATa, NBL2, and D4Z4) were observed (Guo et al., 2014; Hou et al., 2014). A recent study found decreases in DNA methylation of NOS2A (the encoding gene of eNO) and increases in eNO in association with PM2.5 and its major constituents (OC, EC, nitrate and ammonium), suggesting that NOS2A methylation may be important in the pathway of PM2.5-mediated airway inflammation (Chen et al., 2015c). Nevertheless, the underlying mechanisms involved and the health implications of these air pollution–related epigenetic changes have not been fully understood and thus require further investigation.

1.3 EXPOSURES

Accurate measurement on exposure to air pollution is critical for the assessment of associated health effects, as inaccurate measurement could lead to misclassification of exposure and result in uncertainty in the estimated health effects.

1.3.1 Approach for Measurement and Related Methodological Issues

Typical approaches used for exposure measurement in outside environment include proximity, fixed-site air-monitoring, and personal-level measurement. Proximity is a relatively inaccurate estimation method as it assumes that closer proximity to pollution sources (e.g., busy roads) equates greater exposure, which is not always the case. China has established a national routine ambient air-monitoring system for several major air pollutants, including PM_{10}, NO_2, and SO_2, and day-to-day data on these air pollutants have been frequently used in epidemiologic studies (e.g., time-series studies) in China. Epidemiologic studies based on one single city may use air-monitoring data from one site or multiple sites. However, measurement error or misclassification associated with the use of fixed-site air-monitoring data is a major issue encountered by epidemiologic studies (Zeger et al., 2000). An exposure assessment compared the $PM_{2.5}$ and CO exposure concentrations measured at personal levels under several commuting modes (taxi, bus, and cycling) and fixed-site air-monitoring data in a Beijing urban area, and found that exposures measured at personal levels were significantly different from those measured at a fixed site near the study location (Huang et al., 2012a). Generally, nondifferential misclassification of exposure will lead to underestimated health effects associated with air pollution. Interestingly, a recent panel study compared the effects of short-term exposure to outdoor PM obtained from fixed-site air-monitoring stations and outdoor-originated equivalent personal PM (calculated from outdoor PM with incorporation of time-activity data) on lung function in COPD patients, and found that the use of outdoor PM from central air-monitoring stations may overestimate the potential air pollution effects on lung function (Ni et al., 2016). Therefore, research findings based on fixed-site monitoring data should be interpreted cautiously, given the inherent limitation of exposure misclassification and additional efforts are needed to address this issue whenever possible.

Exposure measurement at personal levels could provide more accurate estimates for individuals' "real exposure" to air pollution, and thus help to avoid the issue of exposure misclassification. Instruments are usually placed at or near participants' breathing zone during the measurement periods, which commonly last for a few to 24 hours. However, personal-level exposure measurement is not applicable for large-scale epidemiologic studies due to the measurement burden and economic cost, and only several panel and experimentally designed studies in China have measured air pollution exposures at personal levels (Huang et al., 2012a, 2013; Langrish et al., 2009, 2012; Wu et al., 2010). Studies using personal-level measurement data could provide more reliable estimates for the health effects associated with air pollution exposure for the study participants. Nevertheless, the generalizability of the study findings is often limited, given the limited numbers of participants that could be included in such studies.

Standard measurement protocols with stringent quality controls are essential for reliable air pollution measurement and assessment of associated health effects. This issue is especially important for investigator-self-administered measurement. Standard methods for the measurements of different air pollutants are provided in the modified national AAQS (Ministry of Environmental Protection of China, 2012). Here, we give a brief summary on the measurement methodology for airborne PM in China as an example. The mainstream PM monitoring methods include β-ray method, tapered element oscillating microbalance technology, and gravimetric analysis. However, significant measurement inconsistency may exist for commercially available PM monitors. For example, a previous study demonstrated that $PM_{2.5}$ concentrations measured by three different types of real-time laser light-scattering aerosol monitors varied substantially under the same measurement conditions (Deng et al., 2009b). The measured air pollution levels may also be affected by meteorological factors such as temperature, relative humidity, and wind (Jian et al., 2012). Therefore, it is important to calibrate or control for the potential measurement error in air pollution monitoring. A more reliable technique used for PM measurement is gravimetric analysis, which collects airborne particles on specific filters that are weighed before and after sampling. This technique has been recognized as an official PM measurement method (Ministry of Environmental Protection of China, 2013) and has been increasingly used in investigator-self-administered studies (Huang et al., 2012a; Langrish et al., 2012; Wu et al., 2010, 2013b).

Another approach used for exposure assessment is spatiotemporal air pollution modeling, which can predict general air pollution or individuals' exposure based on routine fixed-site air-monitoring data as well as geographic information, including meteorology, land use and elevation, population density, road networks, and point source emission data (Yanosky et al., 2008). Utility of such modeled data in assessing health effects associated with air pollution has been demonstrated in epidemiologic studies in developed countries (Puett et al., 2011; Beelen et al., 2014). A similar approach has also been applied to model regional air pollution in China (Fu et al., 2009; Zhang et al., 2013; Wu et al., 2015a), whereas further work is needed to link the modeled air pollution data with health observations in China.

In contrast to traditional exposure assessment, which measures the air pollution levels in outside environment, internal exposure assessment estimates the exposure levels by measuring biomarkers in biological samples, which may provide more accurate information regarding the dose of exposure and retention of toxic chemicals. For example, a follow-up study estimated two groups of newborns' exposure to PAHs by measuring PAH-DNA adduct levels in cord blood and found that lower PAH-DNA adduct levels correlated with the reduced levels of ambient PAHs (Tang et al., 2014); the panel study among 120 schoolchildren in China and Korea measured urinary levels of 1-OHP and found a positive association between urinary 1-OHP and MDA levels (Bae et al., 2010); another study measured urinary levels of nine monohydroxylated PAH metabolites among two groups of children from a polluted area near a heavy traffic road and a nonpolluted area near a university campus, and found that children from the polluted area had higher urinary levels of six PAH metabolites than those from the nonpolluted area (Fan et al., 2012). Similar to the personal-level measurement, the biomarker approach for exposure assessment is also not feasible and cost-effective for large-scale studies.

1.3.2 Air Pollutant Levels

As documented in the background, air pollution levels in China are still at the higher end of the world. The typical levels of ambient air pollutants reported in epidemiologic studies in China are generally higher than those reported in studies from developed countries. In a meta-analysis based on data from 33 epidemiologic studies in China, the summarized daily average concentration ranges of major air pollutants were 44–172 μg/m^3 for PM_{10}, 55–177 μg/m^3 for $PM_{2.5}$, 29–113 μg/m^3 for SO_2, 26–70 μg/m^3 for NO_2, and 1.10–1.80 mg/m^3 for CO, and 8-hour average O_3 concentrations ranged in 56–86 μg/m^3 (Shang et al., 2013). Specifically, one previous study measured the ambient PM levels around the Beijing Olympics at a fixed site in Beijing (Peking University) and found that the PM_{10} levels in Beijing were 1.9–3.5 times higher than the other three Olympic cities (Wang et al., 2009b). The mean PM_{10} concentrations during the Olympics were 28.1, 23.7, 44.3, 82.4, and 53.7 μg/m^3 in Atlanta (1996), Sydney (2000), Athens (2004), Peking University (2008), and Beijing average (2008), respectively (Wang et al., 2009b).

PM exposures measured at the personal level appear to be much higher in studies from China as compared with those in studies from developed countries, whereas the differences in levels of gaseous air pollutants are not apparent. For example, one panel study measured 11 taxi drivers' in-car exposure to several air pollutants around the Beijing Olympics, and the average exposure levels were 55.6 μg/m^3 for $PM_{2.5}$, 4.2 μg/m^3 for EC, 2.6 parts per million (ppm) for CO, and 38.2 parts per billion (ppb) for NO_2 (Wu et al., 2011a). In contrast, another panel study in the United States measured 10 patrol troopers' in-car exposure to air pollutants, and the average exposure levels were 23.0 μg/m^3 for $PM_{2.5}$, 2.3 μg/m^3 for EC, 2.6 ppm for CO, and 41.7 ppb for NO_2 (Riediker et al., 2003).

Several studies also reported levels of organic compounds including PAHs and volatile organic compounds (VOCs) in ambient air. One study determined concentrations of 17 carcinogenic PAHs in ambient $PM_{2.5}$ samples during different periods around the Beijing Olympics, and the average total BaP equivalent concentrations of these PAHs changed from 5.95 ng/m^3 in the source-control period (before and during the Olympics) to 11.1 ng/m^3 in non-source-control period (after the Olympics) (Jia et al., 2011b). Another study reported an average BaP equivalent concentration of 82.4 ng/m^3 for 16 PAHs that is much higher than the national AAQS (24-hour average: 2.5 ng/m^3) at the road intersections where the traffic policemen stand during their work time in Tianjin (Hu et al., 2007). In addition, one study measured 27 VOC species in 14 sampling sites in nine southeast coastal cities of China over two 10-day periods in winter and summer, and the average total VOC concentration was 214.3 μg/m^3 in winter and 111.8 μg/m^3 in summer (Tong et al., 2013). Another study reported that the daily average concentration of total benzene homologues (e.g., benzene, toluene, ethylbenzene), the typical components of VOCs, was 11.98 μg/m^3 during summertime in Beijing (Li et al., 2014a).

1.4 RISK ASSESSMENT AND MANAGEMENT

Risk assessment is the determination of quantitative or qualitative value of risk related to a concrete situation and a recognized threat (also called hazard). Ambient air pollution has been recognized as a strong risk factor for chronic respiratory and cardiovascular diseases worldwide, and outdoor air pollution has been classified as a group 1 carcinogen of human cancer (particularly lung cancer) by the International Agency for Research on Cancer. Many risk assessments have been conducted to identify the health risk associated with ambient air pollution in China, among which several assessments also provide quantitative estimates for the potential economic burden that air pollution may cause. To control for the adverse health impact of ambient air pollution, the government has released several modified editions of national AAQS and implemented a series of air pollution control measures.

1.4.1 Sources

Several studies have provided risk assessments for the health impact associated with ambient PM air pollution in China (Jahn et al., 2011; Lai et al., 2014). A previous review based on 13 studies with ambient PM data in the Pearl River Delta (PRD) region estimated the potential preventable premature deaths due to PM air pollution in these cities (Jahn et al., 2011). Almost all PM data during the period 2000–2004 in the PRD region (PM_{10} ranged from 57.4 μg/m^3 in Hong Kong to 88.3 μg/m^3 in Guangzhou) exceeded national and international air quality guidelines, and the estimated impact of current ambient PM_{10} exposure relative to a standard of 40 μg/m^3 causes approximately 117 excess deaths (lower and upper bound: 91, 164) in Guangzhou and 43 excess deaths (lower and upper bound: 32, 62) in Hong Kong per 100,000 population per year. Under the assumption that Guangzhou's about 10 million inhabitants are on average equally burdened by PM, 11,700 (9100–16,400) premature deaths could have been prevented annually if the PM burden was reduced to 40 μg/m^3.

Health risk associated with exposure to organic compounds including PAHs and VOCs is a major focus of air pollution risk assessment in China (Bai et al., 2009; Hu et al., 2007; Jia et al., 2011b; Li et al., 2014a; Xia et al., 2013; Yu et al., 2008; Zhou et al., 2011). PAHs are known for their carcinogenic and mutagenic properties. One risk assessment estimated that the number of lifetime excess cancer cases due to exposure to 17 carcinogenic PAHs measured in ambient $PM_{2.5}$ samples ranged from 6.5 to 518 per million people for the period before and during the Beijing Olympics and from 12.2 to 964 per million people for the period after the Olympics (Jia et al., 2011b). These results suggest a 46% reduction in estimated inhalation cancer risk due to source control measures, suggesting that source control measures during the Beijing Olympics can significantly reduce the inhalation cancer risk associated with PAH exposure in Chinese megacities similar to Beijing. The aforementioned study among policemen in Tianjin also assessed their occupational carcinogenic risk associated with exposure to PAHs, and the estimated risk ranged from 10^{-5} to 10^{-3}, which was much higher than the acceptable risk level of 10^{-6} (one per million) (Hu et al., 2007).

A risk assessment study estimated health risk associated with ambient benzene homologues (e.g., benzene, toluene, ethylbenzene) exposure based on measurement data at a fixed site in Beijing urban area (Li et al., 2014a). The result showed that benzene homologues had no appreciable adverse noncancer health risks for the exposed population, while benzene had a potential carcinogenic risk of 13.4 per million. Another study measured 12 participants' personal and outdoor exposure to VOCs for 5 days and estimated the associated cumulative cancer risk, which were 44 per million for personal exposure and 30 per million for outdoor exposure (Zhou et al., 2011).

A few studies also assessed the health risk associated with airborne heavy metals. For example, one study estimated that the hazard quotient associated with ambient lead exposure was 0.36 for adults and 2.8 for children in a Shanghai district, suggesting that the health risk in children from lead exposure was unacceptable (Chen et al., 2011c). Another study assessed the contamination and health risk of several metal(loid)s (e.g., arsenic, manganese, copper, nickel, zinc, chrome, cadmium, mercury, lead) in outdoor and indoor particles in Guangzhou and found that arsenic was the most risky element in terms of noncarcinogenic risk (Huang et al., 2014a). A recent risk assessment also evaluated the noncarcinogenic risk and carcinogenic risk associated with several metals in $PM_{2.5}$ and PM_{10} in Hangzhou and found that the excess lifetime carcinogenic risks of Cr in particles of five sampling sites were all higher by one or two orders of magnitude than 1 per million, which might pose cancer risks to human health (Niu et al., 2015).

1.4.2 Economic Burden

Ambient air pollution in China may result in considerable economic burden. An earlier study assessed the economic burden related to particulate air pollution in 111 Chinese cities which accounted for 70% national GDP in 2004, and the total annual economic cost caused by PM_{10} was estimated to be approximately USD 29,178.7 million in 2004 (Zhang et al., 2008). Another economic assessment estimated that the implementation of several low-carbon energy scenarios could prevent 2804–8249 and 9870–23,100 ambient PM_{10}-related avoidable deaths in 2010 and 2020, respectively, among Shanghai residents, and the associated economic benefits could reach USD 507–1492 and USD 2642–6192 million, respectively (Chen et al., 2007). A later economic assessment found that population-weighted PM_{10} exposure during the Beijing Olympics came down by 46% and 19% respectively, as compared with the pre- and post-Olympic periods, and the economic cost associated with human health during the Olympics came down by 38% and 16%, respectively, as compared with the pre- and post-Olympics (Hou et al., 2010). In addition, the daily health economic cost associated with ambient PM_{10} ranged from USD 17.12 to 24.52 million during 2005–2008 in Beijing, accounting for 4.61% to 7.27% of the daily GDP in respective years.

1.4.3 Intervention

China's government has implemented a series of intervention measures in order to control the exacerbating ambient air pollution along with the rapid socioeconomic development. These measures include the release of modified national AAQS, implementation of new vehicle emission standards, changes in industrial structure and improved energy efficiency, use of clean energy (e.g., gas, electricity) and preferred use of clean coal, and controlled reduction in pollution sources and emissions. It has been reported that

Beijing's air quality is mainly driven by the interaction between pollution sources change and implementation of air pollution control measures, and the air pollution control has contributed to the reduction in air pollution in Beijing (Zhang et al., 2011a). However, the appearance of haze episodes in recent years indicates that it will still be a long process to clean the ambient air in China, and temporary and less intensive control measures have limited effectiveness in reducing the frequency of hazy weather occurrence (Tao et al., 2014a). Nevertheless, a national monitoring system has been introduced for $PM_{2.5}$, the central government has committed to spend ¥3.4 trillion yuan (~USD 0.55 trillion) on environmental protection in the 12th five-year plan period from 2011 to 2015, and the governmental air pollution prevention action plan has also proposed a series of more stringent control measures (Chinese Central Government, 2013).

1.5 CONCLUSIONS AND RECOMMENDATIONS

A number of studies ranging from epidemiology to toxicology have provided strong evidence that exposure to ambient air pollution has appreciable impact on human health in China. Although the evidence is not always consistent, given the various air pollutants and health endpoints that examined in the scientific investigation and the heterogeneous nature of the differentially designed studies, it is no doubt that ambient air pollution contributes to the increased health burden in the Chinese population. Based on the strength, consistency, and coherence of the existing scientific findings, we summarize the main points that have been achieved regarding ambient air pollution and its health effects in China as follows:

- The ambient levels of major air pollutants (i.e., PM, SO_2, NO_2, CO, and O_3) are generally higher in China than the levels reported in studies from developed countries. Ambient $PM_{2.5}$ concentrations are particularly high in major Chinese cities during recent years.
- A substantial body of epidemiologic evidence supports the relationship between short-term (day-to-day) exposure to ambient air pollution and increased daily mortality. Evidence for the link between short-term exposure and increased morbidity has been accumulating but is less consistent. Cardiovascular and respiratory diseases are two major components that constitute the excess deaths and morbidity events associated with air pollution exposure.
- Several major air pollutants, including PM and gases (SO_2, NO_2, CO, and O_3), all have been associated with the increased adverse health outcomes, although the relative importance between these pollutants has been uncertain, given that it is hard to differentiate the health effects of these pollutants on the exposed population in the context of the natural environment.
- The existing evidence suggests that the air pollution–health risk relationships may be monotonic without a "safe" threshold, though a "flattening out" phenomenon at higher concentrations has been implicated for the air pollution–mortality relationship (Kan et al., 2007).
- Available studies are suggestive that reductions in air pollution levels can improve cardiovascular function among healthy adults and patients within a few hours and days.

- Preliminary evidence suggests that PM's health effects are affected by its physical and chemical properties. Smaller size-fractioned particles and certain chemical constituents of PM (e.g., BC/EC, organics, metals) appear to be more closely related to the observed health effects associated with PM exposure.
- Many biological mechanisms have been proposed for the action of air pollution exposure. Epidemiologic and toxicological evidence has been increasingly supportive for several of the mechanisms, among which increased pulmonary and systemic inflammation and cardiac ANS imbalance have been most frequently documented.

However, the overall evidence for the adverse health impact of ambient air pollution is still underreported in China as compared with that from the developed countries in North America and Europe. A majority of the published studies in China have not been available until recently. Given the high air pollution levels present in China associated with the accelerating socioeconomic development of the nation and the fact that China's population accounts for over 20% of the world's total population, the potential health and economic burdens that could be attributed to exposure to ambient air pollution in China will be tremendous. Therefore, determining the directions for future air pollution health investigation is critical for the prevention of the air pollution–related health burden in China. Based on the lessons learned and shortages present in the available studies, the following strategic areas will be worth while to consider in future investigations in China:

- Investigate the health implications of exposure to the mixture of various air pollutants (e.g., PM and gaseous air pollutants) and differentiate the relative importance of major air pollutants by using integrated research methodologies (e.g., controlled exposure, improved study design, and stringent statistical analysis).
- Better document the influence of PM physical and chemical properties on the observed PM-related health effects and explore the responsible pollution sources behind the observed health effects.
- Better investigate the effective biological mechanisms of pathophysiological relevance (e.g., inflammation, oxidative stress, coagulation) through which short-term and long-term air pollution may promote the development of adverse health outcomes in the context of high air pollution in China. Evidence for the long-term exposure and development of chronic diseases (e.g., hypertension, diabetes, metabolic syndrome, cancer) is especially lacking and needed.
- Integrate the subclinical health effects and adverse health outcomes observed in epidemiologic studies and biological mechanisms of air pollution action elucidated by toxicological studies, and clarify the public health implications of such integrated evidence.
- Determine the long-term effects of ambient air pollution in populations using prospective study design and better exposure assessment approaches. Although the potential long-term effects of air pollution have been implicated by a few epidemiologic studies in China, the risk estimates are diverse and less reliable due to the retrospective study design as well as unperfected exposure assessment methodologies which may lead to substantial exposure misclassification.

The current evidence is far from sufficient for the establishment of exposure–response relationships between mediate (a few to 12 months) to long-term (years) air pollution exposure and adverse health outcomes (e.g., increased incidence and mortality of diseases) in the context of high pollution levels in China. To control for the various biases present in the available studies, future studies using prospective study design (e.g., cohort study) and more reliable exposure assessment methodologies are necessary. Specifically, a prospectively designed cohort study focusing on the pathogenesis of cardiopulmonary diseases associated with particulate air pollution has been established in China and the investigation is underway (Song et al., 2014). Evidence from such long-term follow-up studies will provide important base for the formulation of long-term air quality standards that are practically feasible for China.

- Determine the short-term and long-term health benefits in the population that may be gained owing to the reductions in air pollution levels. As China has strived to control the ambient air pollution with great efforts, pollution levels may vary over short-term periods in the future and gradually be reduced in the long term, and the determination of health benefits associated with pollution reductions will provide evidence for the public health improvements and help to tailor further relevant policies.
- Determine the susceptibility to air pollution in different population subgroups and the interactions between air pollution and traditional health risk factors (e.g., cigarette smoking, alcohol abuse, obesity, dietary factors). Although a few studies have found that certain groups (e.g., elders and poorly educated subjects) are at higher risk of death in response to short-term air pollution exposure, evidence is still sparse in this field. More informative investigation will aid the identification of susceptible populations and provide important implications for the prevention of adverse air pollution effects.
- Compare the exposure–response characteristics of air pollution and health in China with those observed in developed countries in terms of gene–environment interaction. Both genetic, epigenetic, and environmental factors may contribute to the observed air pollution effects (Chen et al., 2015c; Peters et al., 2009; Mordukhovich et al., 2009). However, few studies in China have investigated the potential interactions between air pollution and genetic/epigenetic susceptibility, leaving this topic as an under-investigated area.
- Document the environmental health impact in the context of air pollution–climate change interaction. Climate change is another environmental issue in China and may also increase the health burden (Kan et al., 2012). A few studies have provided preliminary evidence that climate-relevant changes (e.g., temperature) may interact with air pollution to affect the observed health effects (Li et al., 2011; Wu et al., 2013c, 2015b). However, the extent and time course of such interaction have remained largely unclear. China's landscape and related climates are diverse and thus will be an ideal environment for such investigation.
- Assess the health risks of different air pollutants at larger scales (e.g., multicity or national) and provide evidence-based estimates for the health and economic burdens that air pollution may cause in China. Updating the risk

assessments over time may be critical for informative and timely policy making in view that the implementation of air pollution control measures may gradually reduce the air pollution levels over time.

REFERENCES

Aunan, K. and Pan, X. C. (2004). Exposure-response functions for health effects of ambient air pollution applicable for China—A meta-analysis. *Sci Total Environ*, 329, 3–16.

Baccarelli, A., Barretta, F., Dou, C., Zhang, X., Mccracken, J. P., Diaz, A., Bertazzi, P. A., Schwartz, J., Wang, S., and Hou, L. (2011). Effects of particulate air pollution on blood pressure in a highly exposed population in Beijing, China: A repeated-measure study. *Environ Health*, 10, 108.

Baccarelli, A. A., Zheng, Y., Zhang, X., Chang, D., Liu, L., Wolf, K., Zhang, Z., Mccracken, J. P., Diaz, A., Bertazzi, P., Schwartz, J., Wang, S., Kang, C. M., Koutrakis, P., and Hou, L. (2014). Air pollution exposure and lung function in highly exposed subjects in Beijing, China: A repeated-measure study. *Part Fibre Toxicol*, 11, 51.

Badkar, M. (2012). The ultimate guide to China's voracious energy use. *Business Insider.* Available at: http://www.businessinsider.com/china-energy-use-2012-8?op=1. Retrieved 19 February 2014.

Bae, S., Pan, X. C., Kim, S. Y., Park, K., Kim, Y. H., Kim, H., and Hong, Y. C. (2010). Exposures to particulate matter and polycyclic aromatic hydrocarbons and oxidative stress in schoolchildren. *Environ Health Perspect*, 118, 579–83.

Bai, N., Khazaei, M., Van Eeden, S. F., and Laher, I. (2007). The pharmacology of particulate matter air pollution-induced cardiovascular dysfunction. *Pharmacol Ther*, 113, 16–29.

Bai, Z., Hu, Y., Yu, H., Wu, N., and You, Y. (2009). Quantitative health risk assessment of inhalation exposure to polycyclic aromatic hydrocarbons on citizens in Tianjin, China. *Bull Environ Contam Toxicol,* 83, 151–4.

Bao, A., Liang, L., Li, F., Zhang, M., and Zhou, X. (2013). Effects of acute ozone exposure on lung peak allergic inflammation of mice. *Front Biosci (Landmark Ed)*, 18, 838–51.

Beelen, R., Raaschou-Nielsen, O., Stafoggia, M., Andersen, Z. J., Weinmayr, G., Hoffmann, B., Wolf, K., Samoli, E., Fischer, P., Nieuwenhuijsen, M., Vineis, P., Xun, W. W., Katsouyanni, K., Dimakopoulou, K., Oudin, A., Forsberg, B., Modig, L., Havulinna, A. S., Lanki, T., Turunen, A., Oftedal, B., Nystad, W., Nafstad, P., De Faire, U., Pedersen, N. L., Ostenson, C. G., Fratiglioni, L., Penell, J., Korek, M., Pershagen, G., Eriksen, K. T., Overvad, K., Ellermann, T., Eeftens, M., Peeters, P. H., Meliefste, K., Wang, M., Bueno-De-Mesquita, B., Sugiri, D., Kramer, U., Heinrich, J., De Hoogh, K., Key, T., Peters, A., Hampel, R., Concin, H., Nagel, G., Ineichen, A., Schaffner, E., Probst-Hensch, N., Kunzli, N., Schindler, C., Schikowski, T., Adam, M., Phuleria, H., Vilier, A., Clavel-Chapelon, F., Declercq, C., Grioni, S., Krogh, V., Tsai, M. Y., Ricceri, F., Sacerdote, C., Galassi, C., Migliore, E., Ranzi, A., Cesaroni, G., Badaloni, C., Forastiere, F., Tamayo, I., Amiano, P., Dorronsoro, M., Katsoulis, M., Trichopoulou, A., Brunekreef, B., and Hoek, G. (2014). Effects of long-term exposure to air pollution on natural-cause mortality: An analysis of 22 European cohorts within the multicentre ESCAPE project. *Lancet*, 383, 785–95.

Breitner, S., Liu, L., Cyrys, J., Bruske, I., Franck, U., Schlink, U., Leitte, A. M., Herbarth, O., Wiedensohler, A., Wehner, B., Hu, M., Pan, X. C., Wichmann, H. E., and Peters, A. (2011). Sub-micrometer particulate air pollution and cardiovascular mortality in Beijing, China. *Sci Total Environ*, 409, 5196–204.

Byun, H. M., Panni, T., Motta, V., Hou, L., Nordio, F., Apostoli, P., Bertazzi, P. A., and Baccarelli, A. A. (2013). Effects of airborne pollutants on mitochondrial DNA methylation. *Part Fibre Toxicol*, 10, 18.

Cao, J., Li, W., Tan, J., Song, W., Xu, X., Jiang, C., Chen, G., Chen, R., Ma, W., Chen, B., and Kan, H. (2009). Association of ambient air pollution with hospital outpatient and emergency room visits in Shanghai, China. *Sci Total Environ*, 407, 5531–6.

Cao, J., Xu, H., Xu, Q., Chen, B., and Kan, H. (2012). Fine particulate matter constituents and cardiopulmonary mortality in a heavily polluted Chinese city. *Environ Health Perspect*, 120, 373–8.

Cao, J., Yang, C., Li, J., Chen, R., Chen, B., Gu, D., and Kan, H. (2011). Association between long-term exposure to outdoor air pollution and mortality in China: A cohort study. *J Hazard Mater*, 186, 1594–600.

Chen, C., Chen, B., Wang, B., Huang, C., Zhao, J., Dai, Y., and Kan, H. (2007). Low-carbon energy policy and ambient air pollution in Shanghai, China: A health-based economic assessment. *Sci Total Environ*, 373, 13–21.

Chen, J., Tan, M., Nemmar, A., Song, W., Dong, M., Zhang, G., and Li, Y. (2006). Quantification of extrapulmonary translocation of intratracheal-instilled particles *in vivo* in rats: Effect of lipopolysaccharide. *Toxicology*, 222, 195–201.

Chen, R., Chu, C., Tan, J., Cao, J., Song, W., Xu, X., Jiang, C., Ma, W., Yang, C., Chen, B., Gui, Y., and Kan, H. (2010). Ambient air pollution and hospital admission in Shanghai, China. *J Hazard Mater*, 181, 234–40.

Chen, R., Huang, W., Wong, C. M., Wang, Z., Thach, T. Q., Chen, B., and Kan, H. (2012a). Short-term exposure to sulfur dioxide and daily mortality in 17 Chinese cities: The China air pollution and health effects study (CAPES). *Environ Res*, 118, 101–6.

Chen, R., Kan, H., Chen, B., Huang, W., Bai, Z., Song, G., and Pan, G. (2012b). Association of particulate air pollution with daily mortality: The China Air Pollution and Health Effects Study. *Am J Epidemiol*, 175, 1173–81.

Chen, R., Li, Y., Ma, Y., Pan, G., Zeng, G., Xu, X., chen, B., and Kan, H. (2011a). Coarse particles and mortality in three Chinese cities: The China Air Pollution and Health Effects Study (CAPES). *Sci Total Environ*, 409, 4934–8.

Chen, R., Pan, G., Zhang, Y., Xu, Q., Zeng, G., Xu, X., Chen, B., and Kan, H. (2011b). Ambient carbon monoxide and daily mortality in three Chinese cities: The China Air Pollution and Health Effects Study (CAPES). *Sci Total Environ*, 409, 4923–8.

Chen, R., Qiao, L., Li, H., Zhao, Y., Zhang, Y., Xu, W., Wang, C., Wang, H., Zhao, Z., Xu, X., Hu, H., Kan, H. (2015c). Fine particulate matter constituents, nitric oxide synthase DNA methylation and exhaled nitric oxide. *Environ Sci Technol*, 49, 11859–65.

Chen, R., Samoli, E., Wong, C. M., Huang, W., Wang, Z., Chen, B., and Kan, H. (2012c). Associations between short-term exposure to nitrogen dioxide and mortality in 17 Chinese cities: The China Air Pollution and Health Effects Study (CAPES). *Environ Int*, 45, 32–8.

Chen, R., Zhang, Y., Yang, C., Zhao, Z., Xu, X., and Kan, H. (2013a). Acute effect of ambient air pollution on stroke mortality in the China air pollution and health effects study. *Stroke*, 44, 954–60.

Chen, R., Zhao, A., Chen, H., Zhao, Z., Cai, J., Wang, C., Yang, C., Li, H., Xu, X., Ha, S., Li, T., Kan, H. (2015a). Cardiopulmonary benefits of reducing indoor particles of outdoor origin: a randomized, double-blind crossover trial of air purifiers. *J Am Coll Cardiol*, 65, 2279-87.

Chen, R., Zhao, Z., and Kan, H. (2013b). Heavy smog and hospital visits in Beijing, China. *Am J Respir Crit Care Med*, 188, 1170–1.

Chen, T., Jia, G., Wei, Y., and Li, J. (2013c). Beijing ambient particle exposure accelerates atherosclerosis in ApoE knockout mice. *Toxicol Lett*, 223, 146–53.

Chen, Y., Ebenstein, A., Greenstone, M., and Li, H. (2013d). Evidence on the impact of sustained exposure to air pollution on life expectancy from China's Huai River policy. *Proc Natl Acad Sci U S A*, 110, 12936–41.

Chen, Y., Wang, J., Shi, G., Sun, X., Chen, Z., and Xu, S. (2011c). Human health risk assessment of lead pollution in atmospheric deposition in Baoshan District, Shanghai. *Environ Geochem Health*, 33, 515–23.

Chen, R., Zhao, Z., Sun, Q., Lin, Z., Zhao, A., Wang, C., Xia, Y., Xu, X., Kan, H. (2015b). Size-fractionated particulate air pollution and circulating biomarkers of inflammation, coagulation, and vasoconstriction in a panel of young adults. *Epidemiology*, 26, 328-36.

Chinese Central Government (2013). The State Council issued the "Air Pollution Prevention Action Plan" Ten measures (in Chinese). Available at: http://www.gov.cn/jrzg/2013-09/12/content_2486918.htm. Retrieved 31 March 2014.

Deng, F., Guo, X., Liu, H., Fang, X., Yang, M., and Chen, W. (2007). Effects of dust storm PM2.5 on cell proliferation and cell cycle in human lung fibroblasts. *Toxicol In Vitro*, 21, 632–8.

Deng, F. R., Guo, X. B., Hu, J., and Lv, P. (2009a). Acute heart toxicity in rats induced by PM2.5 intratracheal instillation and its mechanisms. *Asian J Ecotox*, 4, 57–62.

Deng, F. R., Guo, X. B., Xia, P. P., and Liu, H. (2010). Comparative Study on Parameters of Blood Coagulation Affected by PM2.5 and Nickel Sulfate Intratracheal Instillation in Rats. *J Environ and Occup Med*, 1, 009.

Deng, F. R., Wang, X., and Wu, S. W. (2009b). Comparative Study of Three Types of Light Scattering Aerosol Monitor. *J Environ Health*, 6, 014.

Deng, X., Zhang, F., Rui, W., Long, F., Wang, L., Feng, Z., Chen, D., and Ding, W. (2013). PM2.5-induced oxidative stress triggers autophagy in human lung epithelial A549 cells. *Toxicol In Vitro*, 27, 1762–70.

Ding, X., Wang, M., Chu, H., Chu, M., Na, T., Wen, Y., Wu, D., Han, B., Bai, Z., Chen, W., Yuan, J., Wu, T., Hu, Z., Zhang, Z., and Shen, H. (2014). Global gene expression profiling of human bronchial epithelial cells exposed to airborne fine particulate matter collected from Wuhan, China. *Toxicol Lett*, 228, 25–33.

Dockery, D. W., Pope, C. A., 3RD, Xu, X., Spengler, J. D., Ware, J. H., Fay, M. E., Ferris, B. G., Jr., and Speizer, F. E. (1993). An association between air pollution and mortality in six U.S. cities. *N Engl J Med*, 329, 1753–9.

Dong, G. H., Qian, Z. M., Xaverius, P. K., Trevathan, E., Maalouf, S., Parker, J., Yang, L., Liu, M. M., Wang, D., Ren, W. H., Ma, W., Wang, J., Zelicoff, A., Fu, Q., and Simckes, M. (2013). Association between long-term air pollution and increased blood pressure and hypertension in China. *Hypertension*, 61, 578–84.

Dong, G.H., Wang, J., Zeng, X.W., Chen, L., Qin, X.D., Zhou, Y., Li, M., Yang, M., Zhao, Y., Ren, W.H., Hu, Q.S. 2015. Interactions between air pollution and obesity on blood pressure and hypertension in Chinese children. *Epidemiology*, 26, 740-7.

Dong, G. H., Zhang, P., Sun, B., Zhang, L., Chen, X., Ma, N., Yu, F., Guo, H., Huang, H., Lee, Y. L., Tang, N., and Chen, J. (2012). Long-term exposure to ambient air pollution and respiratory disease mortality in Shenyang, China: A 12-year population-based retrospective cohort study. *Respiration*, 84, 360–8.

Fan, R., Wang, D., Mao, C., Ou, S., Lian, Z., Huang, S., Lin, Q., Ding, R., and She, J. (2012). Preliminary study of children's exposure to PAHs and its association with 8-hydroxy-2′-deoxyguanosine in Guangzhou, China. *Environ Int*, 42, 53–8.

Fu, J. S., Streets, D. G., Jang, C. J., Hao, J., He, K., Wang, L., and Zhang, Q. (2009). Modeling regional/urban ozone and particulate matter in Beijing, China. *J Air Waste Manag Assoc*, 59, 37–44.

Gao, Y., Chan, E. Y., Li, L. P., He, Q. Q., and Wong, T. W. (2013). Chronic effects of ambient air pollution on lung function among Chinese children. *Arch Dis Child*, 98, 128–35.

Gong, J., Zhu, T., Kipen, H., Wang, G., Hu, M., Guo, Q., Ohman-Strickland, P., Lu, S. E., Wang, Y., Zhu, P., Rich, D. Q., Huang, W., and Zhang, J. (2014). Comparisons of ultrafine and fine particles in their associations with biomarkers reflecting physiological pathways. *Environ Sci Technol.*

Gong, J., Zhu, T., Kipen, H., Wang, G., Hu, M., Ohman-Strickland, P., Lu, S. E., Zhang, L., Wang, Y., Zhu, P., Rich, D. Q., Diehl, S. R., Huang, W., and Zhang, J. J. (2013). Malondialdehyde in exhaled breath condensate and urine as a biomarker of air pollution induced oxidative stress. *J Expo Sci Environ Epidemiol*, 23, 322–7.

Gonzalez-Flecha, B. (2004). Oxidant mechanisms in response to ambient air particles. *Mol Aspects Med*, 25, 169–82.

Guo, L., Byun, H. M., Zhong, J., Motta, V., Barupal, J., Zheng, Y., Dou, C., Zhang, F., Mccracken, J. P., Diaz, A., Marco, S. G., Colicino, S., Schwartz, J., Wang, S., Hou, L., and Baccarelli, A. A. (2014). Effects of short-term exposure to inhalable particulate matter on DNA methylation of tandem repeats. *Environ Mol Mutagen*, 55, 322–35.

Guo, Y., Jia, Y., Pan, X., Liu, L., and Wichmann, H. E. (2009). The association between fine particulate air pollution and hospital emergency room visits for cardiovascular diseases in Beijing, China. *Sci Total Environ*, 407, 4826–30.

Guo, Y., Li, S., Tian, Z., Pan, X., Zhang, J., and Williams, G. (2013). The burden of air pollution on years of life lost in Beijing, China, 2004–08: Retrospective regression analysis of daily deaths. *BMJ*, 347, f7139.

Guo, Y., Tong, S., Li, S., Barnett, A. G., Yu, W., Zhang, Y., and Pan, X. (2010a). Gaseous air pollution and emergency hospital visits for hypertension in Beijing, China: A time-stratified case-crossover study. *Environ Health*, 9, 57.

Guo, Y., Tong, S., Zhang, Y., Barnett, A. G., Jia, Y., and Pan, X. (2010b). The relationship between particulate air pollution and emergency hospital visits for hypertension in Beijing, China. *Sci Total Environ*, 408, 4446–50.

He, M., Ichinose, T., Yoshida, S., Nishikawa, M., Mori, I., Yanagisawa, R., Takano, H., Inoue, K., Sun, G., and Shibamoto, T. (2010a). Airborne Asian sand dust enhances murine lung eosinophilia. *Inhal Toxicol*, 22, 1012–25.

He, Q. Q., Wong, T. W., Du, L., Jiang, Z. Q., Gao, Y., Qiu, H., Liu, W. J., Wu, J. G., Wong, A., and Yu, T. S. (2010b). Effects of ambient air pollution on lung function growth in Chinese schoolchildren. *Respir Med*, 104, 1512–20.

HEI (2010). Outdoor air pollution and health in the developing countries of Asia: A comprehensive review. *Special Report.* Boston, Massachusetts, Health Effects Institute.

Hong, X., Liu, C., Chen, X., Song, Y., Wang, Q., Wang, P., and Hu, D. (2013). Maternal exposure to airborne particulate matter causes postnatal immunological dysfunction in mice offspring. *Toxicology*, 306, 59–67.

Hou, L., Wang, S., Dou, C., Zhang, X., Yu, Y., Zheng, Y., Avula, U., Hoxha, M., Diaz, A., Mccracken, J., Barretta, F., Marinelli, B., Bertazzi, P. A., Schwartz, J., and Baccarelli, A. A. (2012). Air pollution exposure and telomere length in highly exposed subjects in Beijing, China: A repeated-measure study. *Environ Int*, 48, 71–7.

Hou, L., Zhang, X., Dioni, L., Barretta, F., Dou, C., Zheng, Y., Hoxha, M., Bertazzi, P. A., Schwartz, J., Wu, S., Wang, S., and Baccarelli, A. A. (2013). Inhalable particulate matter and mitochondrial DNA copy number in highly exposed individuals in Beijing, China: A repeated-measure study. *Part Fibre Toxicol*, 10, 17.

Hou, L., Zhang, X., Zheng, Y., Wang, S., Dou, C., Guo, L., Byun, H. M., Motta, V., Mccracken, J., Diaz, A., Kang, C. M., Koutrakis, P., Bertazzi, P. A., Li, J., Schwartz, J., and Baccarelli, A. A. (2014). Altered methylation in tandem repeat element and elemental component levels in inhalable air particles. *Environ Mol Mutagen*, 55, 256–65.

Hou, Q., An, X. Q., Wang, Y., and Guo, J. P. (2010). An evaluation of resident exposure to respirable particulate matter and health economic loss in Beijing during Beijing 2008 Olympic Games. *Sci Total Environ*, 408, 4026–32.

Hsiao, W. L., Mo, Z. Y., Fang, M., Shi, X. M., and Wang, F. (2000). Cytotoxicity of PM(2.5) and PM(2.5–10) ambient air pollutants assessed by the MTT and the Comet assays. *Mutat Res*, 471, 45–55.

Hu, Y., Bai, Z., Zhang, L., Wang, X., Zhang, L., Yu, Q., and Zhu, T. (2007). Health risk assessment for traffic policemen exposed to polycyclic aromatic hydrocarbons (PAHs) in Tianjin, China. *Sci Total Environ*, 382, 240–50.

Hua, J., Yin, Y., Peng, L., Du, L., Geng, F., and Zhu, L. (2014). Acute effects of black carbon and PM2.5 on children asthma admissions: A time-series study in a Chinese city. *Sci Total Environ*, 481, 433–8.

Huang, J., Deng, F., Wu, S., and Guo, X. (2012a). Comparisons of personal exposure to PM2.5 and CO by different commuting modes in Beijing, China. *Sci Total Environ*, 425, 52–9.

Huang, J., Deng, F., Wu, S., Lu, H., Hao, Y., and Guo, X. (2013). The impacts of short-term exposure to noise and traffic-related air pollution on heart rate variability in young healthy adults. *J Expo Sci Environ Epidemiol*, 23, 559–64.

Huang, M., Wang, W., Chan, C. Y., Cheung, K. C., Man, Y. B., Wang, X., and Wong, M. H. (2014a). Contamination and risk assessment (based on bioaccessibility via ingestion and inhalation) of metal(loid)s in outdoor and indoor particles from urban centers of Guangzhou, China. *Sci Total Environ*, 479–480, 117–24.

Huang, W., Cao, J., Tao, Y., Dai, L., Lu, S. E., Hou, B., Wang, Z., and Zhu, T. (2012b). Seasonal variation of chemical species associated with short-term mortality effects of PM(2.5) in Xi'an, a central city in China. *Am J Epidemiol*, 175, 556–66.

Huang, W., Wang, G., Lu, S. E., Kipen, H., Wang, Y., Hu, M., Lin, W., Rich, D., Ohman-Strickland, P., Diehl, S. R., Zhu, P., Tong, J., Gong, J., Zhu, T., and Zhang, J. (2012c). Inflammatory and oxidative stress responses of healthy young adults to changes in air quality during the Beijing Olympics. *Am J Respir Crit Care Med*, 186, 1150–9.

Huang, W., Zhu, T., Pan, X., Hu, M., Lu, S. E., Lin, Y., Wang, T., Zhang, Y., and Tang, X. (2012d). Air pollution and autonomic and vascular dysfunction in patients with cardiovascular disease: Interactions of systemic inflammation, overweight, and gender. *Am J Epidemiol*, 176, 117–26.

Huang, Y. B., Song, F. J., Liu, Q., Li, W. Q., Zhang, W., and Chen, K. X. (2014b). A bird's eye view of the air pollution-cancer link in China. *Chin J Cancer*, 33, 176–88.

Ishikawa, H., Tian, Y., Piao, F., Duan, Z., Zhang, Y., Ma, M., Li, H., Yamamoto, H., Matsumoto, Y., Sakai, S., Cui, J., Yamauchi, T., and Yokoyama, K. (2006). Genotoxic damage in female residents exposed to environmental air pollution in Shenyang city, China. *Cancer Lett*, 240, 29–35.

Jahn, H. J., Schneider, A., Breitner, S., Eissner, R., Wendisch, M., and Kramer, A. (2011). Particulate matter pollution in the megacities of the Pearl River Delta, China: A systematic literature review and health risk assessment. *Int J Hyg Environ Health*, 214, 281–95.

Janes, H., Sheppard, L., and Shepherd, K. (2008). Statistical analysis of air pollution panel studies: An illustration. *Ann Epidemiol*, 18, 792–802.

Jia, X., Hao, Y., and Guo, X. (2012a). Ultrafine carbon black disturbs heart rate variability in mice. *Toxicol Lett*, 211, 274–80.

Jia, X., Song, X., Shima, M., Tamura, K., Deng, F., and Guo, X. (2011a). Acute effect of ambient ozone on heart rate variability in healthy elderly subjects. *J Expo Sci Environ Epidemiol*, 21, 541–7.

Jia, X., Song, X., Shima, M., Tamura, K., Deng, F., and Guo, X. (2012b). Effects of fine particulate on heart rate variability in Beijing: A panel study of healthy elderly subjects. *Int Arch Occup Environ Health*, 85, 97–107.

Jia, Y., Stone, D., Wang, W., Schrlau, J., Tao, S., and Simonich, S. L. (2011b). Estimated reduction in cancer risk due to PAH exposures if source control measures during the 2008 Beijing Olympics were sustained. *Environ Health Perspect*, 119, 815–20.

Jia, Y. P., Guo, Y. M., Wang, Z. Y., Xie, Y. Z., Tang, X. Y., Zhu, T., Wang, S., and Pan, X. C. (2009). The correlations between air quality and heart rate variability in aged susceptible people during Beijing Olympic Games 2008. *Zhonghua Yu Fang Yi Xue Za Zhi*, 43, 669–73.

Jia, Y. Q., Zhao, X. H., and Guo, X. B. (2006). Effects of the inhalable particle PM10 on secretion of inflammatory factors in human lung fibroblasts and mouse alveolar macrophage cell. *Wei Sheng Yan Jiu*, 35, 557–60.

Jian, L., Zhao, Y., Zhu, Y. P., Zhang, M. B., and Bertolatti, D. (2012). An application of ARIMA model to predict submicron particle concentrations from meteorological factors at a busy roadside in Hangzhou, China. *Sci Total Environ*, 426, 336–45.

Jiang, L. L., Zhang, Y. H., Song, G. X., Chen, G. H., Chen, B. H., Zhao, N. Q., and Kan, H. D. (2007). A time series analysis of outdoor air pollution and preterm birth in Shanghai, China. *Biomed Environ Sci*, 20, 426–31.

Kan, H., Chen, B., Zhao, N., London, S. J., Song, G., Chen, G., Zhang, Y., and Jiang, L. (2010). Part 1. A time-series study of ambient air pollution and daily mortality in Shanghai, China. *Res Rep Health Eff Inst*, 17–78.

Kan, H., Chen, R., and Tong, S. (2012). Ambient air pollution, climate change, and population health in China. *Environ Int*, 42, 10–9.

Kan, H., Jia, J., and Chen, B. (2004). The association of daily diabetes mortality and outdoor air pollution in Shanghai, China. *J Environ Health*, 67, 21–6.

Kan, H., London, S. J., Chen, G., Zhang, Y., Song, G., Zhao, N., Jiang, L., and Chen, B. (2007). Differentiating the effects of fine and coarse particles on daily mortality in Shanghai, China. *Environ Int*, 33, 376–84.

Kan, H., London, S. J., Chen, G., Zhang, Y., Song, G., Zhao, N., Jiang, L., and Chen, B. (2008). Season, sex, age, and education as modifiers of the effects of outdoor air pollution on daily mortality in Shanghai, China: The Public Health and Air Pollution in Asia (PAPA) Study. *Environ Health Perspect*, 116, 1183–8.

Ko, F. W., Tam, W., Wong, T. W., Chan, D. P., Tung, A. H., Lai, C. K., and Hui, D. S. (2007a). Temporal relationship between air pollutants and hospital admissions for chronic obstructive pulmonary disease in Hong Kong. *Thorax*, 62, 780–5.

Ko, F. W., Tam, W., Wong, T. W., Lai, C. K., Wong, G. W., Leung, T. F., Ng, S. S., and Hui, D. S. (2007b). Effects of air pollution on asthma hospitalization rates in different age groups in Hong Kong. *Clin Exp Allergy*, 37, 1312–9.

Lai, H. K., Tsang, H., Thach, T. Q., and Wong, C. M. (2014). Health impact assessment of exposure to fine particulate matter based on satellite and meteorological information. *Environ Sci Process Impacts*, 16, 239–46.

Lai, H. K., Tsang, H., and Wong, C. M. (2013). Meta-analysis of adverse health effects due to air pollution in Chinese populations. *BMC Public Health*, 13, 360.

Langrish, J. P., Li, X., Wang, S., Lee, M. M., Barnes, G. D., Miller, M. R., Cassee, F. R., Boon, N. A., Donaldson, K., Li, J., Li, L., Mills, N. L., Newby, D. E., and Jiang, L. (2012). Reducing personal exposure to particulate air pollution improves cardiovascular health in patients with coronary heart disease. *Environ Health Perspect*, 120, 367–72.

Langrish, J. P., Mills, N. L., Chan, J. K., Leseman, D. L., Aitken, R. J., Fokkens, P. H., Cassee, F. R., Li, J., Donaldson, K., Newby, D. E., and Jiang, L. (2009). Beneficial cardiovascular effects of reducing exposure to particulate air pollution with a simple facemask. *Part Fibre Toxicol*, 6, 8.

Lee, S. L., Wong, W. H., and Lau, Y. L. (2006). Association between air pollution and asthma admission among children in Hong Kong. *Clin Exp Allergy*, 36, 1138–46.

Leitte, A. M., Schlink, U., Herbarth, O., Wiedensohler, A., Pan, X. C., Hu, M., Richter, M., Wehner, B., Tuch, T., Wu, Z., Yang, M., Liu, L., Breitner, S., Cyrys, J., Peters, A., Wichmann, H. E., and Franck, U. (2011). Size-segregated particle number concentrations and respiratory emergency room visits in Beijing, China. *Environ Health Perspect*, 119, 508–13.

Li, G., Zhou, M., Cai, Y., Zhang, Y., and Pan, X. (2011). Does temperature enhance acute mortality effects of ambient particle pollution in Tianjin City, China. *Sci Total Environ*, 409, 1811–7.

Li, L., Li, H., Zhang, X., Wang, L., Xu, L., Wang, X., Yu, Y., Zhang, Y., and Cao, G. (2014a). Pollution characteristics and health risk assessment of benzene homologues in ambient air in the northeastern urban area of Beijing, China. *J Environ Sci (China)*, 26, 214–23.

Li, P., Zhao, J., Gong, C., Bo, L., Xie, Y., Kan, H., and Song, W. (2014b). Association between individual PM2.5 exposure and DNA damage in traffic policemen. *Journal of Occupational and Environmental Medicine*, 56, e98–e101.

Li, R., Meng, Z., and Xie, J. (2007). Effects of sulfur dioxide on the expressions of MUC5AC and ICAM-1 in airway of asthmatic rats. *Regul Toxicol Pharmacol*, 48, 284–91.

Li, R., Meng, Z., and Xie, J. (2008). Effects of sulfur dioxide on the expressions of EGF, EGFR, and COX-2 in airway of asthmatic rats. *Arch Environ Contam Toxicol*, 54, 748–57.

Li, Y., Wang, W., Kan, H., Xu, X., and Chen, B. (2010). Air quality and outpatient visits for asthma in adults during the 2008 Summer Olympic Games in Beijing. *Sci Total Environ*, 408, 1226–7.

Lin, W., Huang, W., Zhu, T., Hu, M., Brunekreef, B., Zhang, Y., Liu, X., Cheng, H., Gehring, U., Li, C., and Tang, X. (2011). Acute respiratory inflammation in children and black carbon in ambient air before and during the 2008 Beijing Olympics. *Environ Health Perspect*, 119, 1507–12.

Liu, C., Xu, X., Bai, Y., Wang, T. Y., Rao, X., Wang, A., Sun, L., Ying, Z., Gushchina, L., Maiseyeu, A., Morishita, M., Sun, Q., Harkema, J. R., and Rajagopalan, S. (2014a). Air pollution-mediated susceptibility to inflammation and insulin resistance: Influence of CCR2 pathways in mice. *Environ Health Perspect*, 122, 17–26.

Liu, L., Breitner, S., Schneider, A., Cyrys, J., Bruske, I., Franck, U., Schlink, U., Marian Leitte, A., Herbarth, O., Wiedensohler, A., Wehner, B., Pan, X., Wichmann, H. E., and Peters, A. (2013a). Size-fractioned particulate air pollution and cardiovascular emergency room visits in Beijing, China. *Environ Res*, 121, 52–63.

Liu, L. and Zhang, J. (2009). Ambient air pollution and children's lung function in China. *Environ Int*, 35, 178–86.

Liu, M. M., Wang, D., Zhao, Y., Liu, Y. Q., Huang, M. M., Liu, Y., Sun, J., Ren, W. H., Zhao, Y. D., He, Q. C., and Dong, G. H. (2013b). Effects of outdoor and indoor air pollution on respiratory health of Chinese children from 50 kindergartens. *J Epidemiol*, 23, 280–7.

Liu, Q., Baumgartner, J., Zhang, Y., Liu, Y., Sun, Y., and Zhang, M. (2014b). Oxidative potential and inflammatory impacts of source apportioned ambient air pollution in Beijing. *Environ Sci Technol*, 48, 12920–9.

Lu, F., Xu, D., Cheng, Y., Dong, S., Guo, C., Jiang, X., and Zheng, X. (2015). Systematic review and meta-analysis of the adverse health effects of ambient PM2.5 and PM10 pollution in the Chinese population. *Environ Res*, 136, 196–204.

Lu, X. L., Zhang, X. R., Deng, F. R., and Guo, X. B. (2011). Systemic oxidative stress induced by intratracheal instilling with PM10 in rats. *Beijing Da Xue Xue Bao*, 43, 352–5.

Madaniyazi, L., Guo, Y., Ye, X., Kim, D., Zhang, Y., and Pan, X. (2013). Effects of airborne metals on lung function in inner Mongolian schoolchildren. *J Occup Environ Med*, 55, 80–6.

McKinsey Global Institute (2009). Preparing for China's urban billion. Available at: http://www.mckinsey.com/~/media/McKinsey/dotcom/Insights%20and%20pubs/MGI/Research/Urbanization/Preparing%20for%20Chinas%20urban%20billion/MGI_Preparing_for_Chinas_Urban_Billion_full_report.ashx. Retrieved 19 February 2014.

Meng, X., Ma, Y., Chen, R., Zhou, Z., Chen, B., and Kan, H. (2013). Size-fractionated particle number concentrations and daily mortality in a Chinese city. *Environ Health Perspect*, 121, 1174–8.

Ministry of Environmental Protection of China (2012). Ambient Air Quality Standards (AAQS) (in Chinese). Available at: http://kjs.mep.gov.cn/hjbhbz/bzwb/dqhjbh/dqhjzlbz/201203/W020120410330232398521.pdf. Retrieved 21 March 2014, Ministry of Environmental Protection of the People's Republic of China.

Ministry of Environmental Protection of China (2013). Specifications and test procedures for PM10 and PM2.5 sampler (in Chinese). Available at: http://kjs.mep.gov.cn/hjbhbz/bzwb/dqhjbh/jcgfffbz/201308/W020130802491940182117.pdf. Retrieved 31 2014, Ministry of Environmental Protection of the People's Republic of China.

Ministry of Environmental Protection of China (2014). 2013 Environment in China (in Chinese). Available at: http://jcs.mep.gov.cn/hjzl/zkgb/. Retrieved 14 Feb 2015, Ministry of Environmental Protection of the People's Republic of China.

Mordukhovich, I., Wilker, E., Suh, H., Wright, R., Sparrow, D., Vokonas, P. S., and Schwartz, J. (2009). Black carbon exposure, oxidative stress genes, and blood pressure in a repeated-measures study. *Environ Health Perspect*, 117, 1767–72.

National Bureau of Statistics of China (2015). National Bureau of Statistics of the People's Republic of China (1996–2014). Available at: http://www.stats.gov.cn/tjsj/ndsj/. Retrieved 14 Feb 2015.

Ni, Y., Wu, S., Ji, W., Chen, Y., Zhao, B., Shi, S., Tu, X., Li, H., Pan, L., Deng, F., Guo, X. (2016). The exposure metric choices have significant impact on the association between short-term exposure to outdoor particulate matter and changes in lung function: Findings from a panel study in chronic obstructive pulmonary disease patients. *Sci Total Environ*, 542, 264-70.

Niu, L., Ye, H., Xu, C., Yao, Y., and Liu, W. (2015). Highly time- and size-resolved fingerprint analysis and risk assessment of airborne elements in a megacity in the Yangtze River Delta, China. *Chemosphere*, 119, 112–21.

Pan, G., Zhang, S., Feng, Y., Takahashi, K., Kagawa, J., Yu, L., Wang, P., Liu, M., Liu, Q., Hou, S., Pan, B., and Li, J. (2010). Air pollution and children's respiratory symptoms in six cities of Northern China. *Respir Med*, 104, 1903–11.

Peng, R. D., Chang, H. H., Bell, M. L., Mcdermott, A., Zeger, S. L., Samet, J. M., and Dominici, F. (2008). Coarse particulate matter air pollution and hospital admissions for cardiovascular and respiratory diseases among Medicare patients. *JAMA*, 299, 2172–9.

Peng, Z., Yu, S., Zhang, Z., Liu, G., He, L., Liao, X., Zhang, L., Wu, H., and Wu, Y. (2011). Effect of ambient air PM10 concentration on the hospital outpatient visit of respiratory diseases in Shenzhen City. *Wei Sheng Yan Jiu*, 40, 485–8.

Peters, A., Greven, S., Heid, I. M., Baldari, F., Breitner, S., Bellander, T., Chrysohoou, C., Illig, T., Jacquemin, B., Koenig, W., Lanki, T., Nyberg, F., Pekkanen, J., Pistelli, R., Ruckerl, R., Stefanadis, C., Schneider, A., Sunyer, J., and Wichmann, H. E. (2009). Fibrinogen genes modify the fibrinogen response to ambient particulate matter. *Am J Respir Crit Care Med*, 179, 484–91.

Pope, C. A., III, Burnett, R. T., Thurston, G. D., Thun, M. J., Calle, E. E., Krewski, D., and Godleski, J. J. (2004). Cardiovascular mortality and long-term exposure to particulate air pollution: Epidemiological evidence of general pathophysiological pathways of disease. *Circulation*, 109, 71–7.

Pope, C. A., III, Burnett, R. T., Turner, M. C., Cohen, A., Krewski, D., Jerrett, M., Gapstur, S. M., and Thun, M. J. (2011). Lung cancer and cardiovascular disease mortality associated with ambient air pollution and cigarette smoke: Shape of the exposure-response relationships. *Environ Health Perspect*, 119, 1616–21.

Pope, C. A., III, and Dockery, D. W. (2006). Health effects of fine particulate air pollution: Lines that connect. *J Air Waste Manag Assoc*, 56, 709–42.

Puett, R. C., Hart, J. E., Schwartz, J., Hu, F. B., Liese, A. D., and Laden, F. (2011). Are particulate matter exposures associated with risk of type 2 diabetes? *Environ Health Perspect*, 119, 384–9.

Qian, Z., Lin, H. M., Stewart, W. F., Kong, L., Xu, F., Zhou, D., Zhu, Z., Liang, S., Chen, W., Shah, N., Stetter, C., and He, Q. (2010). Seasonal pattern of the acute mortality effects of air pollution. *J Air Waste Manag Assoc*, 60, 481–8.

Qiao, L., Cai, J., Wang, H., Wang, W., Zhou, M., Lou, S., Chen, R., Dai, H., Chen, C., and Kan, H. (2014). PM2.5 constituents and hospital emergency-room visits in Shanghai, China. *Environ Sci Technol*, 48, 10406–14.

Qiu, H., Yu, I. T., Wang, X., Tian, L., Tse, L. A., and Wong, T. W. (2013). Cool and dry weather enhances the effects of air pollution on emergency IHD hospital admissions. *Int J Cardiol*, 168, 500–5.

Rich, D. Q., Kipen, H. M., Huang, W., Wang, G., Wang, Y., Zhu, P., Ohman-Strickland, P., Hu, M., Philipp, C., Diehl, S. R., Lu, S. E., Tong, J., Gong, J., Thomas, D., Zhu, T., and Zhang, J. J. (2012). Association between changes in air pollution levels during the

Beijing Olympics and biomarkers of inflammation and thrombosis in healthy young adults. *JAMA*, 307, 2068–78.
Riediker, M., Williams, R., Devlin, R., Griggs, T., and Bromberg, P. (2003). Exposure to particulate matter, volatile organic compounds, and other air pollutants inside patrol cars. *Environ Sci Technol*, 37, 2084–93.
Samet, J. M., Dominici, F., Curriero, F. C., Coursac, I., and Zeger, S. L. (2000). Fine particulate air pollution and mortality in 20 U.S. cities, 1987–1994. *N Engl J Med*, 343, 1742–9.
Sang, N., Yun, Y., Yao, G. Y., Li, H. Y., Guo, L., and Li, G. K. (2011). SO2-induced neurotoxicity is mediated by cyclooxygenases-2-derived prostaglandin E(2) and its downstream signaling pathway in rat hippocampal neurons. *Toxicol Sci*, 124, 400–13.
Shang, Y., Sun, Z., Cao, J., Wang, X., Zhong, L., Bi, X., Li, H., Liu, W., Zhu, T., and Huang, W. (2013). Systematic review of Chinese studies of short-term exposure to air pollution and daily mortality. *Environ Int*, 54, 100–11.
Song, Y., Hou, J., Huang, X., Zhang, X., Tan, A., Rong, Y., Sun, H., Zhou, Y., Cui, X., Yang, Y., Guo, Y., Zhang, Z., Luo, X., Zhang, B., Hou, F., He, X., Xie, J., Wu, T., Chen, W., Yuan, J. (2014). The Wuhan-Zhuhai (WHZH) cohort study of environmental air particulate matter and the pathogenesis of cardiopulmonary diseases: study design, methods and baseline characteristics of the cohort. *BMC Public Health*, 14, 994.
Sun, Y., Song, X., Han, Y., Ji, Y., Gao, S., Shang, Y., Lu, S.E., Zhu, T., Huang, W. (2015). Size-fractioned ultrafine particles and black carbon associated with autonomic dysfunction in subjects with diabetes or impaired glucose tolerance in Shanghai, China. *Part Fibre Toxicol*, 12, 8.
Tam, W. W., Wong, T. W., Ng, L., Wong, S. Y., Kung, K. K., and Wong, A. H. (2014). Association between air pollution and general outpatient clinic consultations for upper respiratory tract infections in Hong Kong. *PLoS One*, 9, e86913.
Tam, W. W., Wong, T. W., and Wong, A. H. (2012a). Effect of dust storm events on daily emergency admissions for cardiovascular diseases. *Circ J*, 76, 655–60.
Tam, W. W., Wong, T. W., Wong, A. H., and Hui, D. S. (2012b). Effect of dust storm events on daily emergency admissions for respiratory diseases. *Respirology*, 17, 143–8.
Tang, D., Li, T. Y., Chow, J. C., Kulkarni, S. U., Watson, J. G., Ho, S. S., Quan, Z. Y., Qu, L. R., and Perera, F. (2014). Air pollution effects on fetal and child development: A cohort comparison in China. *Environ Pollut*, 185, 90–6.
Tao, J., Zhang, L., Zhang, Z., Huang, R., Wu, Y., Zhang, R., Cao, J., and Zhang, Y. (2014a). Control of PM2.5 in Guangzhou during the 16th Asian Games period: Implication for hazy weather prevention. *Sci Total Environ*, 508, 57–66.
Tao, Y., Huang, W., Huang, X., Zhong, L., Lu, S. E., Li, Y., Dai, L., Zhang, Y., and Zhu, T. (2012). Estimated acute effects of ambient ozone and nitrogen dioxide on mortality in the Pearl River Delta of southern China. *Environ Health Perspect*, 120, 393–8.
Tao, Y., Mi, S., Zhou, S., Wang, S., and Xie, X. (2014b). Air pollution and hospital admissions for respiratory diseases in Lanzhou, China. *Environ Pollut*, 185, 196–201.
Tao, Y., Zhong, L., Huang, X., Lu, S. E., Li, Y., Dai, L., Zhang, Y., Zhu, T., and Huang, W. (2011). Acute mortality effects of carbon monoxide in the Pearl River Delta of China. *Sci Total Environ*, 410–411, 34–40.
Tian, L., Qiu, H., Pun, V. C., Lin, H., Ge, E., Chan, J. C., Louie, P. K., Ho, K. F., and Yu, I. T. (2013). Ambient carbon monoxide associated with reduced risk of hospital admissions for respiratory tract infections. *Am J Respir Crit Care Med*, 188, 1240–5.
Tie, X., Wu, D., and Brasseur, G. (2009). Lung cancer mortality and exposure to atmospheric aerosol particles in Guangzhou, China. *Atmospheric Environment*, 43, 2375–2377.
Tong, L., Liao, X., Chen, J., Xiao, H., Xu, L., Zhang, F., Niu, Z., and Yu, J. (2013). Pollution characteristics of ambient volatile organic compounds (VOCs) in the southeast coastal cities of China. *Environ Sci Pollut Res Int*, 20, 2603–15.

USEPA (2012). National Ambient Air Quality Standards (NAAQS). Available at: http://www.epa.gov/air/criteria.html. Retrieved 23 March 2014, United States Environmental Protection Agency.

Valavanidis, A., Fiotakis, K., and Vlachogianni, T. (2008). Airborne particulate matter and human health: Toxicological assessment and importance of size and composition of particles for oxidative damage and carcinogenic mechanisms. *J Environ Sci Health C Environ Carcinog Ecotoxicol Rev*, 26, 339–62.

Van Donkelaar, A., Martin, R. V., Brauer, M., Kahn, R., Levy, R., Verduzco, C., and Villeneuve, P. J. (2010). Global estimates of ambient fine particulate matter concentrations from satellite-based aerosol optical depth: Development and application. *Environ Health Perspect*, 118, 847–55.

Venners, S. A., Wang, B., Xu, Z., Schlatter, Y., Wang, L., and Xu, X. (2003). Particulate matter, sulfur dioxide, and daily mortality in Chongqing, China. *Environ Health Perspect*, 111, 562–7.

Wang, F. F., Zheng, C. J., and Guo, X. B. (2006). Effect of PM2.5 collected during the dust and non-dust periods on the viability and gap junctional intercellular communication in human lung fibroblasts. *Wei Sheng Yan Jiu*, 35, 26–30.

Wang, S., Zhang, J., Zeng, X., Zeng, Y., Wang, S., and Chen, S. (2009a). Association of traffic-related air pollution with children's neurobehavioral functions in Quanzhou, China. *Environ Health Perspect*, 117, 1612–8.

Wang, W., Primbs, T., Tao, S., and Simonich, S. L. (2009b). Atmospheric particulate matter pollution during the 2008 Beijing Olympics. *Environ Sci Technol*, 43, 5314–20.

Wang, X., Chen, R., Meng, X., Geng, F., Wang, C., and Kan, H. (2013). Associations between fine particle, coarse particle, black carbon and hospital visits in a Chinese city. *Sci Total Environ*, 458–460, 1–6.

Wang, X., Deng, F. R., Lv, H. B., Wu, S. W., and Guo, X. B. (2011). Long-term effects of air pollution on the occurrence of respiratory symptoms in adults of Beijing. *Beijing Da Xue Xue Bao*, 43, 356–9.

Wang, X., Deng, F. R., Wu, S. W., Zheng, Y. D., Sun, X. M., Liu, H., and Guo, X. B. (2010). Short-time effects of inhalable particles and fine particles on children's lung function in a district in Beijing. *Beijing Da Xue Xue Bao*, 42, 340–4.

Wang, X., Ding, H., Ryan, L., and Xu, X. (1997). Association between air pollution and low birth weight: A community-based study. *Environ Health Perspect*, 105, 514–20.

Wang, X., Westerdahl, D., Chen, L. C., Wu, Y., Hao, J., Pan, X., Guo, X., and Zhang, K. M. (2009c). Evaluating the air quality impacts of the 2008 Beijing Olympic Games: On-road emission factors and black carbon profiles. *Atmospheric Environment*, 43, 4535–4543.

Wei, A. and Meng, Z. (2006). Induction of chromosome aberrations in cultured human lymphocytes treated with sand dust storm fine particles (PM2.5). *Toxicol Lett*, 166, 37–43.

Wei, Y., Han, I. K., Shao, M., Hu, M., Zhang, O. J., and Tang, X. (2009). PM2.5 constituents and oxidative DNA damage in humans. *Environ Sci Technol*, 43, 4757–62.

Wei, Y., Han, I. K., Hu, M., Shao, M., Zhang, J. J., and Tang, X. (2010). Personal exposure to particulate PAHs and anthraquinone and oxidative DNA damages in humans. *Chemosphere*, 81, 1280–5.

Wong, C. M., Atkinson, R. W., Anderson, H. R., Hedley, A. J., Ma, S., Chau, P. Y., and Lam, T. H. (2002a). A tale of two cities: Effects of air pollution on hospital admissions in Hong Kong and London compared. *Environ Health Perspect*, 110, 67–77.

Wong, C. M., Thach, T. Q., Chau, P. Y., Chan, E. K., Chung, R. Y., Ou, C. Q., Yang, L., Peiris, J. S., Thomas, G. N., Lam, T. H., Wong, T. W., and Hedley, A. J. (2010). Part 4. Interaction between air pollution and respiratory viruses: Time-series study of daily mortality and hospital admissions in Hong Kong. *Res Rep Health Eff Inst*, 283–362.

Wong, C. M., Vichit-Vadakan, N., Kan, H., and Qian, Z. (2008). Public Health and Air Pollution in Asia (PAPA): A multicity study of short-term effects of air pollution on mortality. *Environ Health Perspect*, 116, 1195–202.

Wong, T. W., Lau, T. S., Yu, T. S., Neller, A., Wong, S. L., Tam, W., and Pang, S. W. (1999). Air pollution and hospital admissions for respiratory and cardiovascular diseases in Hong Kong. *Occup Environ Med*, 56, 679–83.

Wong, T. W., Tam, W. S., Yu, T. S., and Wong, A. H. (2002b). Associations between daily mortalities from respiratory and cardiovascular diseases and air pollution in Hong Kong, China. *Occup Environ Med*, 59, 30–5.

World Health Organization (2006). Air Quality Guidelines Global Update 2005: Particulate matter, ozone, nitrogen dioxide and sulfur dioxide. Copenhagen: WHO Regional Office for Europe.

World Health Organization (2012). Exposure to particulate matter with an aerodynamic diameter of 10 um or less (PM10) in 1100 urban areas, 2003–2010.

Wu, S., Deng, F., Hao, Y., Shima, M., Wang, X., Zheng, C., Wei, H., Lv, H., Lu, X., Huang, J., Qin, Y., and Guo, X. (2013a). Chemical constituents of fine particulate air pollution and pulmonary function in healthy adults: The Healthy Volunteer Natural Relocation study. *J Hazard Mater*, 260, 183–91.

Wu, S., Deng, F., Huang, J., Wang, H., Shima, M., Wang, X., Qin, Y., Zheng, C., Wei, H., Hao, Y., Lv, H., Lu, X., and Guo, X. (2013b). Blood pressure changes and chemical constituents of particulate air pollution: Results from the healthy volunteer natural relocation (HVNR) study. *Environ Health Perspect*, 121, 66–72.

Wu, S., Deng, F., Hao, Y., Wang, X., Zheng, C., Lv, H., Lu, X., Wei, H., Huang, J., Qin, Y., Shima, M., and Guo, X. (2014b). Fine particulate matter, temperature, and lung function in healthy adults: Findings from the HVNR study. *Chemosphere*.

Wu, S., Deng, F., Liu, Y., Shima, M., Niu, J., Huang, Q., and Guo, X. (2013c). Temperature, traffic-related air pollution, and heart rate variability in a panel of healthy adults. *Environ Res*, 120, 82–9.

Wu, S., Deng, F., Niu, J., Huang, Q., Liu, Y., and Guo, X. (2010). Association of heart rate variability in taxi drivers with marked changes in particulate air pollution in Beijing in 2008. *Environ Health Perspect*, 118, 87–91.

Wu, S., Deng, F., Niu, J., Huang, Q., Liu, Y., and Guo, X. (2011a). Exposures to PM2.5 components and heart rate variability in taxi drivers around the Beijing 2008 Olympic Games. *Sci Total Environ*, 409, 2478–85.

Wu, S., Deng, F., Niu, J., Huang, Q., Liu, Y., and Guo, X. (2011b). The relationship between traffic-related air pollutants and cardiac autonomic function in a panel of healthy adults: A further analysis with existing data. *Inhal Toxicol*, 23, 289–303.

Wu, S., Deng, F., Wang, X., Wei, H., Shima, M., Huang, J., Lv, H., Hao, Y., Zheng, C., Qin, Y., and Guo, X. (2013d). Association of lung function in a panel of young healthy adults with various chemical components of ambient fine particulate air pollution in Beijing, China. *Atmospheric Environment*, 77, 873–884.

Wu, S., Deng, F., Wei, H., Huang, J., Wang, H., Shima, M., Wang, X., Qin, Y., Zheng, C., Hao, Y., and Guo, X. (2012). Chemical constituents of ambient particulate air pollution and biomarkers of inflammation, coagulation and homocysteine in healthy adults: A prospective panel study. *Part Fibre Toxicol*, 9, 49.

Wu, S., Deng, F., Wei, H., Huang, J., Wang, X., Hao, Y., Zheng, C., Qin, Y., Lv, H., Shima, M., and Guo, X. (2014c). Association of cardiopulmonary health effects with source-appointed ambient fine particulate in Beijing, China: A combined analysis from the Healthy Volunteer Natural Relocation (HVNR) study. *Environ Sci Technol*, 48, 3438–48.

Wu, J., Li, J., Peng, J., Li, W., Xu, G., and Dong, C. (2015a). Applying land use regression model to estimate spatial variation of PM2.5 in Beijing, China. *Environ Sci Pollut Res Int*, 22, 7045-61.

Wu, S., Deng, F., Huang, J., Wang, X., Qin, Y., Zheng, C., Wei, H., Shima, M., Guo, X. (2015b). Does ambient temperature interact with air pollution to alter blood pressure? A repeated-measure study in healthy adults. *J Hypertens*, 33, 2414-21.

Wu, S., Yang, D., Wei, H., Wang, B., Huang, J., Li, H., Shima, M., Deng, F., Guo, X. (2015c). Association of chemical constituents and pollution sources of ambient fine particulate air pollution and biomarkers of oxidative stress associated with atherosclerosis: A panel study among young adults in Beijing, China. *Chemosphere*, 135, 347-53.

Xia, Z., Duan, X., Tao, S., Qiu, W., Liu, D., Wang, Y., Wei, S., Wang, B., Jiang, Q., Lu, B., Song, Y., and Hu, X. (2013). Pollution level, inhalation exposure and lung cancer risk of ambient atmospheric polycyclic aromatic hydrocarbons (PAHs) in Taiyuan, China. *Environ Pollut*, 173, 150–6.

Xu, M., Guo, Y., Zhang, Y., Westerdahl, D., Mo, Y., Liang, F., and Pan, X. (2014). Spatiotemporal analysis of particulate air pollution and ischemic heart disease mortality in Beijing, China. *Environmental Health*, 13, 109.

Xu, M. M., Jia, Y. P., Li, G. X., Liu, L. Q., Mo, Y. Z., Jin, X. B., and Pan, X. C. (2013). Relationship between ambient fine particles and ventricular repolarization changes and heart rate variability of elderly people with heart disease in Beijing, China. *Biomed Environ Sci*, 26, 629–37.

Xu, X., Deng, F., Guo, X., Lv, P., Zhong, M., Liu, C., Wang, A., Tzan, K., Jiang, S. Y., Lippmann, M., Rajagopalan, S., Qu, Q., Chen, L. C., and Sun, Q. (2012). Association of systemic inflammation with marked changes in particulate air pollution in Beijing in 2008. *Toxicol Lett*, 212, 147–56.

Xu, X., Li, B., and Huang, H. (1995). Air pollution and unscheduled hospital outpatient and emergency room visits. *Environ Health Perspect*, 103, 286–9.

Xu, Z., Xu, X., Zhong, M., Hotchkiss, I. P., Lewandowski, R. P., Wagner, J. G., Bramble, L. A., Yang, Y., Wang, A., Harkema, J. R., Lippmann, M., Rajagopalan, S., Chen, L. C., and Sun, Q. (2011). Ambient particulate air pollution induces oxidative stress and alterations of mitochondria and gene expression in brown and white adipose tissues. *Part Fibre Toxicol*, 8, 20.

Yang, C., Chen, A., Chen, R., Qi, Y., Ye, J., Li, S., Li, W., Liang, Z., Liang, Q., Guo, D., Kan, H., and Chen, X. (2014). Acute effect of ambient air pollution on heart failure in Guangzhou, China. *International Journal of Cardiology*, 177, 436–41.

Yang, G., Wang, Y., Zeng, Y., Gao, G. F., Liang, X., Zhou, M., Wan, X., Yu, S., Jiang, Y., Naghavi, M., Vos, T., Wang, H., Lopez, A. D., and Murray, C. J. (2013). Rapid health transition in China, 1990–2010: Findings from the Global Burden of Disease Study 2010. *Lancet*, 381, 1987–2015.

Yanosky, J. D., Paciorek, C. J., Schwartz, J., Laden, F., Puett, R., and Suh, H. H. (2008). Spatio-temporal modeling of chronic PM10 exposure for the Nurses' Health Study. *Atmos Environ (1994)*, 42, 4047–62.

Yoshida, T., Yoshioka, Y., Fujimura, M., Yamashita, K., Higashisaka, K., Nakanishi, R., Morishita, Y., Kayamuro, H., Nabeshi, H., Nagano, K., Abe, Y., Kamada, H., Tsunoda, S., Yoshikawa, T., Itoh, N., and Tsutsumi, Y. (2010). Potential adjuvant effect of intranasal urban aerosols in mice through induction of dendritic cell maturation. *Toxicol Lett*, 199, 383–8.

Yu, Y., Guo, H., Liu, Y., Huang, K., Wang, Z., and Zhan, X. (2008). Mixed uncertainty analysis of polycyclic aromatic hydrocarbon inhalation and risk assessment in ambient air of Beijing. *J Environ Sci (China)*, 20, 505–12.

Yuan, Z., Chen, Y., Zhang, Y., Liu, H., Liu, Q., Zhao, J., Hu, M., Huang, W., Wang, G., Zhu, T., Zhang, J., and Zhu, P. (2013). Changes of plasma vWF level in response to the improvement of air quality: An observation of 114 healthy young adults. *Ann Hematol*, 92, 543–8.

Zeger, S. L., Thomas, D., Dominici, F., Samet, J. M., Schwartz, J., Dockery, D., and Cohen, A. (2000). Exposure measurement error in time-series studies of air pollution: Concepts and consequences. *Environ Health Perspect*, 108, 419–26.

Zhang, H., Liu, Y., Shi, R., and Yao, Q. (2013). Evaluation of PM10 forecasting based on the artificial neural network model and intake fraction in an urban area: A case study in Taiyuan City, China. *J Air Waste Manag Assoc*, 63, 755–63.

Zhang, J., Ouyang, Z., Miao, H., and Wang, X. (2011a). Ambient air quality trends and driving factor analysis in Beijing, 1983–2007. *J Environ Sci (China)*, 23, 2019–28.

Zhang, J., Song, H., Tong, S., Li, L., Liu, B., and Wan, L. (2000). Ambient sulfate concentration and chronic disease mortality in Beijing. *Sci Total Environ*, 262, 63–71.

Zhang, J. J., Hu, W., Wei, F., Wu, G., Korn, L. R., and Chapman, R. S. (2002). Children's respiratory morbidity prevalence in relation to air pollution in four Chinese cities. *Environ Health Perspect*, 110, 961–7.

Zhang, L. W., Chen, X., Xue, X. D., Sun, M., Han, B., Li, C. P., Ma, J., Yu, H., Sun, Z. R., Zhao, L. J., Zhao, B. X., Liu, Y. M., Chen, J., Wang, P. P., Bai, Z. P., and Tang, N. J. (2014). Long-term exposure to high particulate matter pollution and cardiovascular mortality: A 12-year cohort study in four cities in northern China. *Environ Int*, 62, 41–7.

Zhang, M., Song, Y., Cai, X., and Zhou, J. (2008). Economic assessment of the health effects related to particulate matter pollution in 111 Chinese cities by using economic burden of disease analysis. *J Environ Manage*, 88, 947–54.

Zhang, P., Dong, G., Sun, B., Zhang, L., Chen, X., Ma, N., Yu, F., Guo, H., Huang, H., Lee, Y. L., Tang, N., and Chen, J. (2011b). Long-term exposure to ambient air pollution and mortality due to cardiovascular disease and cerebrovascular disease in Shenyang, China. *PLoS One*, 6, e20827.

Zhang, Y. H., Chen, C. H., Chen, G. H., Song, G. X., Chen, B. H., Fu, Q. Y., and Kan, H. D. (2006). Application of DALYs in measuring health effect of ambient air pollution: A case study in Shanghai, China. *Biomed Environ Sci*, 19, 268–72.

Zhao, A., Chen, R., Kuang, X., and Kan, H. (2014a). Ambient air pollution and daily outpatient visits for cardiac arrhythmia in Shanghai, China. *Journal of Epidemiology*, 24, 321–6.

Zhao, A., Chen, R., Wang, C., Zhao, Z., Yang, C., Lu, J., Chen, X., Kan, H. (2015). Associations between size-fractionated particulate air pollution and blood pressure in a panel of type II diabetes mellitus patients. *Environ Int*, 80, 19-25.

Zhao, J., Gao, Z., Tian, Z., Xie, Y., Xin, F., Jiang, R., Kan, H., and Song, W. (2013). The biological effects of individual-level PM2.5 exposure on systemic immunity and inflammatory response in traffic policemen. *Occup Environ Med*, 70, 426–31.

Zhao, X., Sun, Z., Ruan, Y., Yan, J., Mukherjee, B., Yang, F., Duan, F., Sun, L., Liang, R., Lian, H., Zhang, S., Fang, Q., Gu, D., Brook, J. R., Sun, Q., Brook, R. D., Rajagopalan, S., and Fan, Z. (2014b). Personal black carbon exposure influences ambulatory blood pressure: Air pollution and cardiometabolic disease (AIRCMD-China) study. *Hypertension*, 63, 871–7.

Zhou, J., You, Y., Bai, Z., Hu, Y., Zhang, J., and Zhang, N. (2011). Health risk assessment of personal inhalation exposure to volatile organic compounds in Tianjin, China. *Sci Total Environ*, 409, 452–9.

Zhou, M., He, G., Fan, M., Wang, Z., Liu, Y., Ma, J., Ma, Z., Liu, J., Liu, Y., Wang, L., and Liu, Y. (2015a). Smog episodes, fine particulate pollution and mortality in China. *Environ Res*, 136, 396–404.

Zhou, M., He, G., Liu, Y., Yin, P., Li, Y., Kan, H., Fan, M., and Xue, A. (2015b). The associations between ambient air pollution and adult respiratory mortality in 32 major Chinese cities, 2006–2010. *Environ Res*, 137C, 278–286.

Zhou, M., Liu, Y., Wang, L., Kuang, X., Xu, X., and Kan, H. (2014a). Particulate air pollution and mortality in a cohort of Chinese men. *Environ Pollut*, 186, 1–6.

Zhou, N., Cui, Z., Yang, S., Han, X., Chen, G., Zhou, Z., Zhai, C., Ma, M., Li, L., Cai, M., Li, Y., Ao, L., Shu, W., Liu, J., and Cao, J. (2014b). Air pollution and decreased semen quality: A comparative study of Chongqing urban and rural areas. *Environ Pollut*, 187, 145–52.

2 Air Pollution and Health in Taiwan

Chang-Chuan Chan, Chia-Pin Chio, and Szu-Ying Chen

CONTENTS

2.1 INTRODUCTION

Along with the rapid industrialization and motorization, various mobile and stationary emission sources contribute high level of air pollution. The health impacts of ambient air pollution began to being noticed since the year of 1980s in Europe and North America. Numerous epidemiological studies in Western countries have provided strong evidences on the hazardous effects of ambient air pollution on human beings, especially in cardiopulmonary diseases (Anderson et al., 2012; Brook et al., 2010; Carlsten and Georas, 2014). However, the study results from Western countries may not completely reflect the health burden of air pollution in Asian countries because of disparities in air pollution levels and constituents, population structure, disease patterns, lifestyle, and environmental stressors. In the past decades, Asian countries experienced enormous economic development and rapid transitions in urbanization, motorization, and industrialization, which was accompanied by high air pollution levels. In addition to urban air pollution, the long-ranged transboundary air pollution, such as Asian dust storm (ADS) or smoke haze, from Southeast Asia contributes additional health impacts. As a result, some Asian countries, including Japan, South Korea, China, Taiwan, and Hong-Kong (China), became aware of this issue and conducted many studies on the health impacts of air pollution. Since the first Taiwan Environmental Protection Administration's air quality monitoring (AQM) station was officially operated in 1993, there have been more than 150 studies in evaluation of health effects of air pollution published in peer-reviewed English journals. Of these scientific papers, approximately 80% addressed the cardiorespiratory effects with short-term exposures to air pollution. Others investigated the effects of air pollution on central nervous system, allergic responses, reproductive and developmental diseases, and cancers. This chapter reviews, summarizes, and integrates the evidence of relationships between exposures to air pollution and cardiopulmonary health endpoints from studies in the past 20 years in Taiwan, as in Figure 2.1.

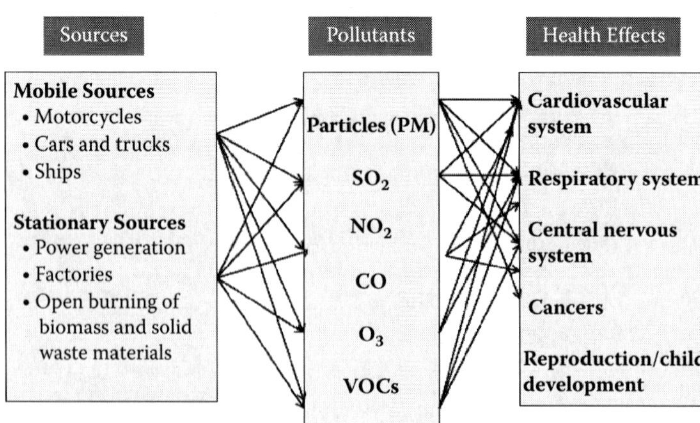

FIGURE 2.1 The framework describing the relationships among source, pollutants, and health effects.

2.2. HEALTH EFFECTS

2.2.1 Epidemiological Studies

In Taiwan, the case-crossover, time series, and cross-sectional designed cohort studies were conducted in epidemiological studies. Diseases of cardiovascular, central nervous, and respiratory systems are of the most concern. More details, such as their relevant biological mechanisms of toxicity, study design, outcome of measurements, and so on, are described in the following sections.

2.2.2 Cardiovascular System

Large numbers of epidemiological studies in Taiwan have consistently demonstrated the effects of short-term exposures to particulate matter (PM) and gaseous pollutants on a variety of cardiovascular effects, from morbidity or mortality of cardiovascular disease (CVD) to subtle clinical manifestations (e.g., blood pressure [BP] change). The possible biological pathways linking PM and CVD also have been well evaluated in serial panel and experimental studies. More detailed descriptions for each study are discussed in the following sections.

2.2.2.1 Cardiovascular Mortality

Four epidemiological studies either with case-crossover design or time-series design consistently demonstrated a positive association between PM and CVD (Figure 2.2). Two case-crossover studies conducted in southern and northern Taiwan consistently reported $PM_{2.5}$ was associated to increased risk of death from circulatory disease with risk estimates 1.44 (95% confidence interval [CI]: 1.21–1.71) for an interquartile range (IQR) of $PM_{2.5}$ (39.4 μg/m^3) in southern Taiwan and 1.07 (95% CI: 1.01–1.15) for an IQR of $PM_{2.5}$ (17.2 μg/m^3) in northern Taiwan, respectively (Tsai et al., 2014b, 2014a). Another two time-series studies also found that cardiovascular mortality was associated with PM_{10} in central Taiwan, with risk estimates of 1.041 (95% CI: 1.015–1.068) and 1.12 (95% CI: 0.998–1.258) for increments of 46.4 μg/m^3 and 40.0 μg/m^{3}of PM_{10}, respectively (Tsai et al., 2010; Liang et al., 2009). The effect estimates of cardiovascular mortality are especially higher in Kaohsiung City, which has the

Study	Location	Exposure Increment (μg/m^3)	Covariates
$PM_{2.5}$			
Tsai et al. (2014)	Taipei	17.2	All age, winter
Tsai et al. (2014)	Kaohsiung	38.4	All age, winter
PM_{10}			
Linag et al. (2009)	Taichung	40.0	65+
Tsai et al. (2010)	Taichung	46.4	All age

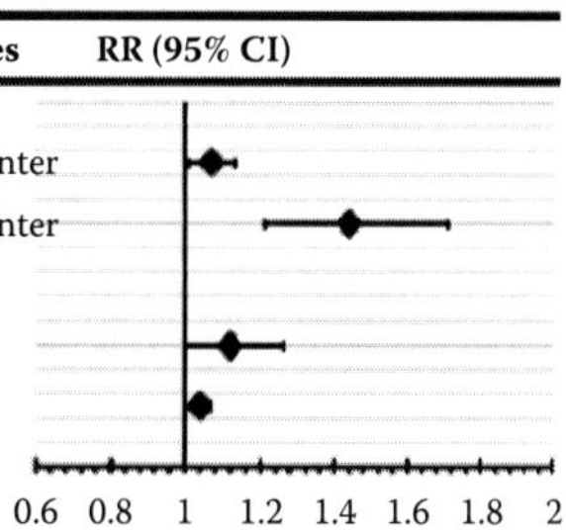

FIGURE 2.2 Summary of risk estimates of mortality for cardiovascular disease with short-term exposures to $PM_{2.5}$ and PM_{10} in Taiwan. Studies presented in this figure were conducted with either case-crossover or time-series design.

largest heavy and petrochemical industrial complex in Taiwan. The geographical variance may indicate a constituent-specific or concentration–response effects on relationship of PM-related cardiovascular mortality.

In addition to PM, Chan and Ng (2011) demonstrated that the long-range transported (LRT) ADS is also associated with the excess deaths for cardiovascular events (odds ratio: 1.045 [95% CI: 1.001–1.081]) among populations above 65 years old.

2.2.2.2 Hospital Admissions and Emergency Department Visits

More than 10 epidemiological studies either with case-crossover design or time-series design consistently showed an excess risk of hospital admissions or emergency department visits for a variety of cardiovascular events with short-term exposures to PM (Figure 2.3). Serial case-crossover studies using the National Health Insurance research database demonstrate the short-term exposures to PM_{10} are associated with increased hospital admissions or emergency room visits for a variety of CVD, including congestive heart failure, cardiac arrhythmia, and acute myocardial infarction (Hsieh et al., 2010; Tsai et al., 2009; Chiu and Yang, 2009; Cheng et al., 2009b; Yang, 2008; Lee et al., 2007b; Chang et al., 2005; Yang et al., 2004). $PM_{2.5}$ is also associated with hospital admissions or emergency room visits for congestive heart failure, cardiac arrhythmia, and acute myocardial infarction (Hsieh et al., 2013; Chiu et al., 2013; Chang et al., 2013). Still, the risk estimates of cardiovascular hospital admissions in response to short-term exposures to PM are significantly higher in Kaohsiung than in other cities. The study results also indicated that seasonality may modify the relationships between PM and CVD morbidity (Chang et al., 2005, 2013; Chiu et al., 2013; Hsieh et al., 2010, 2013; Yang et al., 2004; Yang, 2008; Cheng et al., 2009b).

Study	Location	Outcome	Exposure Increment (μg/m³)	Covariates
$PM_{2.5}$				
Chang et al. (2013)	Taipei	AMI	17.5	All age, warm season
Chiu et al. (2013)	Taipei	Arrhythmia	17.5	All age, warm season
Hsieh et al. (2013)	Taipei	CHF	17.5	All age, warm season
PM_{10}				
Yang et al. (2004)	Kaohsiung	CVD	66.3	All age, cold season
Chang et al. (2005)	Taipei	CVD	24.5	All age, cold season
Lee et al. (2008)	Taipei	CHF	27.5	All age
Yang et al. (2008)	Taipei	CHF	27	All age, warm season
Cheng et al. (2009)	Kaohsiung	AMI	61.9	All age, cold season
Hsieh et al. (2010)	Taipei	AMI	27.2	All age, warm season
Tsai et al. (2012)	Taipei	AMI	27.3	All age
Chang et al. (2013)	Taipei	AMI	27.5	All age, warm season

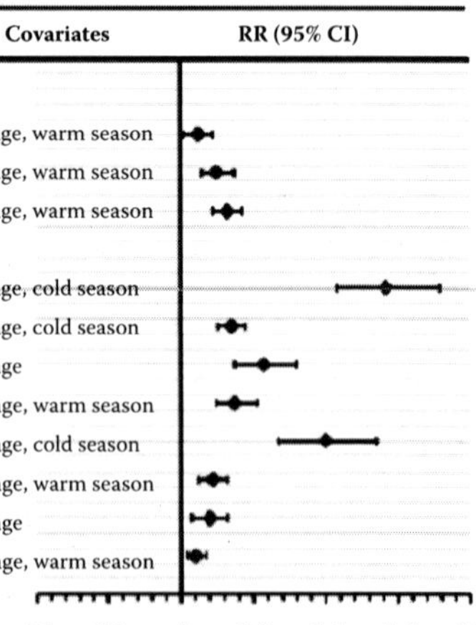

FIGURE 2.3 Summary of risk estimates of hospital admissions or emergency room visits for cardiovascular disease with short-term exposures to $PM_{2.5}$ and PM_{10} in Taiwan. Studies presented in this figure were conducted with either case-crossover or time-series design.

2.2.2.3 Biological Pathway

There has been substantial improvement in our understanding of the biological mechanisms involved in PM-mediated cardiovascular effects. The American Heart Association scientific statements in 2010 proposed the plausible biological pathways linking PM inhalation and extrapulmonary effects on the cardiovascular system (Brook et al., 2010). These include autonomic nerve system imbalance, systemic inflammation and oxidative stress, and the translocation of PM or particle constituents (organic compounds, metals) into the systemic circulation and breaking the integrity of vasculature. In Taiwan, serial panel studies were also conducted to elucidate the possible biological pathways in relation to PM and CVD.

2.2.2.4 Systemic Inflammation, Oxidative Stress, and Thrombogenicity

A panel study recruited 49 patients with coronary heart disease or with two or more cardiovascular risk factors revealed that high air pollution days with PM_{10} concentration greater than 100 $\mu g/m^3$ during 8–18 hours could increase plasma levels of plasminogen activator inhibitor-1 (PAI-1) compared with that in low PM_{10} days (Su et al., 2006). Another panel study on 76 healthy and young adults in Taiwan demonstrated that PM_{10}, $PM_{2.5}$, and PM components of sulfate and nitrate were associated with increases in high-sensitivity C-reactive protein, fibrinogen, and PAI-1 (Chuang et al., 2007). The 8-hydroxy-2′-deoxyguanosine, an oxidative stress marker, was also observed in association with PM fraction size and components (Chuang et al., 2007).

2.2.2.5 Heart Rate Variability

A panel study with nine young subjects in Taipei showed that 1–4-hour moving average number concentrations of submicron particles with a size range of 0.02–1 μm ($NC_{0.02-1}$) were associated with decreases in both time- and frequency-domain heart rate variability (HRV) indices (Chan et al., 2004). This is the first study to examine the effects of personal exposure to ultrafine particles on HRV. Another Taiwanese study also observed a time-domain HRV reduction with exposures to PM with diameters between 0.3 and 1.0 μm ($PM_{0.3-1}$) in cardiac and hypertensive populations (Chuang et al., 2005). Chuang et al. (2007) further found that, in addition to $PM_{2.5}$, the components of $PM_{2.5}$, such as sulfate and organic carbon (OC), were associated with HRV reduction in 46 patients with risk for CVD.

2.2.2.6 Blood Pressure

One of the important air pollution–mediated alternations on cardiovascular outcomes is the variation in BP. However, the results from both epidemiological and experimental studies are still inconclusive on PM-related changes in BP. In Taiwan, two experimental studies were conducted to evaluate the changes in BP with exposures of concentrated particles (Cheng et al., 2003; Chang et al., 2004). Cheng et al. (2003) found that the hourly mean BP decreased in pulmonary hypertensive rats after the particle exposure. Contrarily, Chang et al. (2004) reported the maximum increase of mean BP up to 8.7 mmHg in spontaneous hypertensive rats at the end of exposure to concentrated $PM_{2.5}$. Chan et al. (2004) conducted a panel study and found that with 1–3 hours of $NC_{0.02-1}$ exposure, systolic and diastolic BP significantly increased in 10 patients with lung function impairment. The population-based study conducted by Chuang et al. (2010) observed slightly elevated systolic BP with exposures to PM_{10} among 7578 subjects. However, another population-based study with 9238 nonsmoking adults found that both PM_{10} and gaseous pollutants, including sulfur dioxide (SO_2), nitrogen dioxide (NO_2), carbon monoxide (CO), and ozone (O_3), were all associated with decreases in systolic BP and pulse pressure (Chen et al., 2012). This study also found that the estimates of reduction in pulse pressure with exposures to PM_{10} were stronger among men, persons more than 60 years of age, those with hypertension, and those living in the industrial township (Chen et al., 2012). Chen et al. (2014a) further found that 10 μg/m^3 increase in $PM_{2.5}$ was associated with 1.0 mmHg (95% CI: 0.2–1.8 mmHg) narrowing in the pulse pressure, but not systolic and diastolic BP, among nondippers (subjects with nocturnal BP dip of <10% and in high CVD risk). The inconsistent results may indicate that short-term BP variation in response to PM is a complicated physiological response and is tightly regulated by numerous cardiac and vascular homeostatic mechanisms, which are worth more investigation. Also, individual susceptibility or particulate constituents may modify the BP variation in response to air pollution.

2.2.2.7 Cardiac Contractility

Two animal studies observed decreased left ventricular fractional shortening or the maximum rate of left ventricular pressure rise (LV dP/dt max), which indicated cardiac systolic or diastolic dysfunction, with exposure to short-term exposure to diesel exhaust particles (DEPs) in rats (Huang et al., 2010; Yan et al., 2008). Chen et al. (2014a) conducted a panel study with 161 health subjects by 24-hour continuous hemodynamic monitoring and demonstrated that nondippers had a decrease in LV dP/dt max to 3.1% (95% CI: 1.4%–4.8%) for a 10 μg/m^3 increase in $PM_{2.5}$.

2.2.2.8 Vasomotor Function

One recent panel study in Taiwan reported that an IQR increase in personal exposures to O_3 or $PM_{1.0-2.5}$ was associated with a 4.8% and 2.5% increase, respectively, in cardio-ankle vascular index, a surrogate marker of vascular tone, among young, healthy mail carriers (Wu et al., 2010). The panel study by Chen et al. (2014a) used Dynapulse to continuously monitor hemodynamic parameters and directly demonstrated the systemic vascular resistance (SVR) increased 3.6% (95% CI: 1.6%–5.7%) for a 10 μg/m^3 increase in $PM_{2.5}$ among 79 nondippers.

2.2.3 Central Nervous System

Cerebrovascular disease (CBVD) is one of the leading causes of mortality, especially for East Asian populations, accounting for 10%–16% of total deaths. The proportions of cerebral hemorrhage, small artery lacunar infarct, and intracranial atherosclerosis are more common in Asian populations than in white populations. Studies in Western countries have not yet provided the sufficient evidence on the relationship between PM and CBVD; however, serial epidemiological studies in

Study	Location	Exposure increment (μg/m³)	Covariates
$PM_{2.5}$			
Chan et al. (2006)	Taipei	19.7	All age
Chiu et al. (2013)	Taipei	17.5	All age, warm season
Chen et al. (2014)	Taipei	18.5	65+, warm season
PM_{10}			
Tsai et al. (2003)	Kaohsiung	66.3	All age, warm season
Chan et al. (2006)	Taipei	25.4	All age

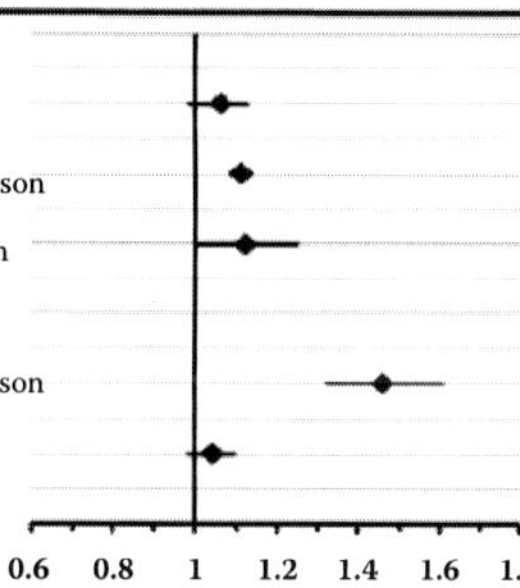

FIGURE 2.4 Summary of risk estimates of hospital admissions or emergency room visits for ischemic stroke with short-term exposures to $PM_{2.5}$ and PM_{10} in Taiwan. Studies presented in this figure were conducted with either case-crossover or time-series design.

Taiwan have demonstrated there is prominent ethnic difference in air pollution–related CBVD between Asian and white populations (Figure 2.4). Chan et al. (2006) conducted a time-series study and found that $PM_{2.5}$ was positively associated with increased emergency room visits for CBVD. Chen et al. (2014b) further demonstrated that $PM_{2.5}$ chemical components, nitrate and elemental carbon, were associated with increased risk of emergency room visits for hemorrhagic stroke, while $PM_{2.5}$ and its chemical component, OC, were associated with increased risk of emergency visits for ischemic stroke in patients aged 65 years or older and female patients. Two case-crossover studies also confirmed the positive associations between stroke admissions and short-term exposures to PM_{10} and $PM_{2.5}$ (Chiu and Yang, 2013; Tsai et al., 2003a). The effect of PM on stroke morbidity was found to be stronger in the warm season (Chiu and Yang, 2013; Tsai et al., 2003a; Chen et al., 2014a).

ADS was also demonstrated to associate with stroke admissions. Yang et al. (2005) conducted a case-crossover study and found the risk of intracranial hemorrhagic stroke admissions was increased 3 days after ADS events. Kang et al. (2013) reported that the numbers of ischemic stroke admissions were significantly higher 1- to 2-days after ADS events.

2.2.4 Respiratory System

The effects of short-term exposures to PM on respiratory outcomes were most studied and addressed worldwide. In Taiwan, a large number of epidemiological and experimental studies have provided strong evidence on the hazardous effects of short-term PM exposures on the respiratory system, including decrease in lung function and increase in morbidity of pulmonary diseases. However, the current studies in Taiwan have not yet provided sufficient evidence to demonstrate the relationship between respiratory mortality and PM. The findings for the associations of various respiratory outcomes and short-term exposures to PM are described in further detail below.

2.2.4.1 Respiratory Mortality

Three case-crossover studies and one time-series study were conducted to evaluate the associations of respiratory-related mortality and short-term exposures to PM (Tsai et al., 2003b, 2014a, 2014b; Liang et al., 2009). However, the results all failed to find the significantly positive effects of PM_{10} or $PM_{2.5}$ on the mortality of respiratory diseases. Another case-crossover study also did not find the excess death for respiratory deaths during ADS (Chan and Ng, 2011). Only an epidemiological study conducted by Chen et al. (2004) reported that the effects of ADS increased 7.66% in the risk for respiratory deaths 1 day after the ADS event.

2.2.4.2 Respiratory Morbidity

A number of cross-sectional studies were conducted at the end of the 20th century to evaluate the associations between air pollution and respiratory symptoms and asthma prevalence among children and adolescents. Using the nationwide questionnaire survey, these study results found positive associations between air pollution and respiratory symptoms (e.g., cough, chronic cough, shortness of breath, and nasal symptoms) and asthma prevalence among children and adolescents (Chen et al., 1998, 1999; Guo et al., 1999; Wang et al., 1999; Lin et al., 2001). Some epidemiological studies also reported an exacerbation of allergic rhinitis following exposures to ambient air pollution (Lee et al., 2003; Yu et al., 2005; Hwang et al., 2006). Serial case-crossover studies using National Health Insurance research database in Taiwan further demonstrated that PM_{10} was associated with both of outpatient and inpatient visits of respiratory diseases, including asthma, chronic obstructive pulmonary disease (COPD), and pneumonia (Tsai et al., 2006; Cheng et al., 2007, 2009a; Lee et al., 2007a; Yang et al., 2007; Yang and Chen, 2007; Chiu et al., 2009). Two time-series studies also found the positive associations between PM_{10} and the outpatient visits for COPD, asthma, and pneumonia (Wang and Chau, 2013; Pan et al., 2014). $PM_{2.5}$ was also demonstrated to associate with hospital admissions for pneumonia, COPD, and asthma (Tang et al., 2007; Tsai et al., 2013; Tsai and Yang, 2014). In addition to PM, four epidemiological studies found that the respiratory clinic visits and hospital admissions for COPD and pneumonia were significantly higher during and after high dust events (Chiu et al., 2008; Chan et al., 2008; Kang et al., 2013; Yu et al., 2013).

2.2.4.3 Pulmonary Function

The associations between pulmonary function and exposures to urban air pollution have been widely surveyed among different study populations. One cross-sectional study observed the inversely associations between PM_{10} and forced vital capacity (FVC) among adolescents (Chang et al., 2012). Others found that gaseous pollutants, including O_3, and nitrogen oxides (NOx) are demonstrated to associate with the reduction of a variety of pulmonary function parameters, including FVC, forced expiratory volume in one second (FEV_1), and peak expiratory flow rate (PEFR) among school-aged children (Chen et al., 1999; Lee et al., 2011; Chang et al., 2012). One panel study conducted by Chan and Wu (2005) demonstrated that the night PEFR for 43 mail carriers was maximally decreased by 0.69% for 10-ppb increase in the 8-hour average O_3 concentration at a 1-day lag.

2.3. EXPOSURES

2.3.1 Approach for Measurement

In Taiwan, the air pollutants data are obtained from two approaches. One is from an active queue management (AQM) network by using automatic instruments maintained by Taiwan Environment Protection Administration (TEPA), and the other one is from academy researches by using manual sampling methods. Hourly based data can be obtained and showed geographically from the website (http://taqm.epa.gov.tw/taqm/tw/default.aspx). The principles for installing the AQM stations should take into account many reasons such as station types, source distributions and categories, pollutants distributions, geographical and meteorological conditions, population distributions and traffic status, and urban, regional, and other land-use plans. Now, 79 AQM stations are online, including 60, 5, 6, 2, 4, and 2 for general, industrial, traffic, national park, background, and other purposes, respectively. In addition, nine fixed stations and two mobile trucks are installed for photochemical purposes.

In the general AQM station, beta-gauge method is always used for monitoring PM_{10} and $PM_{2.5}$; however, ultraviolet (UV) fluorescence and chemiluminescence methods are used for monitoring the SOx and NOx levels. UV absorption method and nondispersive infrared spectroscopy (NDIR) are used to analyze the O_3 and CO, respectively. These regular ambient AQM instruments installed by TEPA, including VEREWA-F701, Met-One BAM-1020, Ecotech 9850, Ecotech 9841, Ecotech 9810, and Horiba for automatically monitoring the PM10, PM2.5, SOx, NOx, O3, and CO, respectively. Recently, R&P 1400 was added for monitoring the 24-hour PM2.5 level each three days by using manually weighting method. In photochemical AQM stations, however, 54 major precursors of ozone production can be obtained by using gas chromatography with flame ionization detector (GC-FID).

For manual sampling methods, high-volume and dichotomous samplers were selected in early researches. The former one were measuring total suspended particles (TSP) or PM_{10} with over 1 cubic meters air sampling per minute; however, the latter one was always used to measure finer two size fractions such as $PM_{2.5}$ and $PM_{2.5-10}$ with 16.7 liters per minute. In fact, the particle-size distributions from any microenvironment or any season are also concerned. The particle-size distribution generally uses 8-stages cascade or 10-stages micro-orifice uniform deposition impactor (MOUDI). The personal environmental monitor (PEM) and Harvard impactor (HI), however, are currently selected to measure the personal $PM_{2.5}$ or PM_{10} exposures. The above manual sampling methods are based on the impact mechanism of particles in the collected filter. Specially, the dichotomous has a virtual impactor to split two streams on the filters and collect two size fractions of particles (Chio et al., 2004, 2014b; Lee et al., 2010).

2.3.2 Air Pollutant Levels

Figure 2.5 shows the averaged air pollutants levels of PM_{10}, $PM_{2.5}$, SO_2, NO_2, O_3, and CO monitored from the AQM stations during 1999–2013. Annual PM_{10} levels ranged from 51.2 μg/m^3 (in 2012) to 63.2 μg/m^3 (in 2005); however, annual $PM_{2.5}$ levels

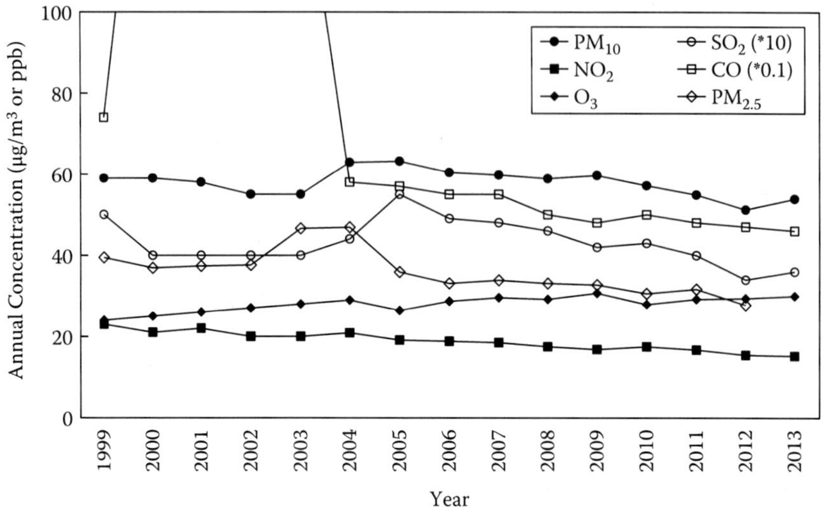

FIGURE 2.5 Major monitoring pollutants in Taiwan during 1999–2013.

ranged from 27.7 μg/m^3 (in 2012) to 46.9 μg/m^3 (in 2004). They showed the trend since 2005. At the same time, the annual SO_2 levels ranged from 3.4 ppb (in 2012) to 5.5 ppb (in 2005) and had same decay trends like PM. Annual NO_2 levels showed a decreasing trend since 1999 with peak levels of 23 ppb (in 1999) and minimum value of 15.2 ppb (in 2013). Basically, the annual CO levels were also shown a decay trend since 2004. The annual O_3 levels showed an inverse trend compared with other pollutants, the values ranged from 24 ppb (in 1999) to 30.7 ppb (in 2009). General speaking, most air pollutants levels are decreasing each year, except for O_3.

2.3.3 Biomarkers

Recently, Cheng et al (2001), Pan et al. (2008), Hu et al. (2011), and Yuan et al. (2014) showed the biomarkers measured such as metals, 1-hydroxypyrene (1-OHP), thiodiglycolic acid (TdGA), and so on in urine or blood. Blood and urinary metals can be monitored for long- and short-term exposures of metals from ingestion, inhalation, and dermal contact pathways to such pollutants emitted from industrial and nonindustrial sources by using inductively coupled plasma mass spectrometry (ICP-MS). For example, vanadium in urine can be considered as a good biomarker emitted from oil burning/refinery industrial source. Urinary 1-OHP is a suggested biomarker for polycyclic aromatic hydrocarbons (PAHs) metabolite by using high-performance liquid chromatography with postcolumn fluorescence derivatization (HPLC/FLD). The urinary TdGA is the major metabolite and a suggested biomarker of vinyl chloride monomer (VCM) emissions. In addition, Lee et al. (2010) showed that urinary 8-hydroxydeoxyquanosine (8-OHdG) level is a biomarker of cellular oxidative stress for exposure to ambient $PM_{2.5}$ or occupational DEP.

2.4. RISK ASSESSMENT AND MANAGEMENT

2.4.1 Sources

In Taiwan, many sources are concerned because of their large emissions. The following sections describe these sources and their emitted pollutants, including PM, O_3, SO_2, NOx, CO, CO_2, volatile organic compounds (VOCs), heavy metals, and so on.

2.4.1.1 Coal-Fired Power Plant

In central Taiwan, the Taichung Power Plant (TCP) is considered as the world's largest coal-fired power plant, with a full designed capacity of 5780 MW (Taiwan Power Company [TPC], 2011; Kuo et al., 2014). The Carbon Monitoring for Action (CARMA, 2014) reported that the TCP was the largest CO_2 contributor in the world, with about 36×10^6 ton per year in 2014. This amount of CO_2 is equivalent 0.1% and 13% emissions in the world and Taiwan. Even when electrostatic precipitators (ESP), flue gas desulfurization (FGD), and selective catalyst reduction (SCR) control devices were installed, over 14,000 and 25,000 ton per year of SO_2 and NOx still emitted from TCP in 2007 (CTCI, 2009).

2.4.1.2 No. 6 Naphtha Cracking Complex

The No. 6 Naphtha Cracking Complex, located on the west coast of central Taiwan (Mailiao Township, Yunlin County), has an overall developed area of 2603 ha. This complex contains 64 plants and 381 stacks, including oil refineries, naphtha cracking plants, cogeneration plants, coal-fired power plants, heavy machinery plants, boiler plants, and downstream petrochemical-related plants (FPCC, 2012; Shie et al., 2013; Shie and Chan, 2013). Six categories of the SOx emission sources are listed, including cogeneration plants (2549 ton/yr, 41.0% of overall complex), coal-fired power plant (1803 ton/yr, 29.0%), oil refineries (654 ton/yr, 10.5%), aromatics and plastics plants (493 ton/yr, 7.9%), naphtha cracking plants (403 ton/yr, 6.5%), and downstream petrochemical plants (314 ton/yr, 5.1%) in 2009 (Shie et al., 2013). Not only petrochemical manufacturing processes but also power generation facilities in the complex contribute to the major SOx emissions. In 2010, a report shows the annual air pollutants emitted by the complex were 3,340 tons, 16,000 tons, 19,622 tons, 4,302 tons for PM, SO_2, NOx, and VOCs, respectively (Formosa Plastic Group, 2010; Chen et al., 2014d).

2.4.1.3 Asian Dust Storm

Taiwan is located at the center of the western Asia Pacific Rim by geographic point of view. Yet the climate change of Taiwan is strongly correlated with the monsoon from Eastern Asia areas, such as Mainland China, Mongolia, and Siberia, especially for spring season every year. Based on the mesoscale meteorology, the LRT aerosols (or ADS) have been evidenced to exacerbate the air quality and impact the human health in Taiwan. The possible pollutants include PM, viruses, and fungi (Cheng et al. 2005; Lee et al. 2006; Chao et al. 2012). On the other hand, many adverse outcomes and mortality are also found to be associated with these ADS events (Lei et al. 2004; Chan et al. 2008).

Recently, several approaches are developed to study these impacts of air quality and human health. Except for monitoring methods, various mechanistic and statistical model approaches are also reviewed here. Both qualitative and quantitative analyses can obtain a better solution for evaluating the impacts from the LRT aerosols.

During ADS periods, about 3%–12% and 25%–150% impact of PM_{10} mass concentrations for annually and maximum monthly, respectively, are determined by analyzing the AQM station. Precisely, the concentrations of coarse particles ($PM_{2.5-10}$) and their crustal elements are significantly elevated (Liu and Shiu 2001). In the epidemiological approach, results show that the cardiovascular death for residents of age greater than 65 may rise 4.5% (95% CI: 1.1%–8.1%) with analyzing multiyears datasets during the ADS periods (Chan and Ng 2011).

According to the aforementioned reviews, the air quality and human health in Taiwan had evidenced to be impacted by the LRT aerosols from Mainland China, Mongolia, and Siberia. Obviously, the major reason of ADS event came from the monsoon in spring every year, and then the coarse particles ($PM_{2.5-10}$) might have twofold increased. Reviews also showed that the susceptible elder groups might have death risk with CVDs. Therefore, the appropriate protections for the susceptible groups should be proposed as soon as possible.

2.4.2 Modeling and Risk Assessment

Several modeling approaches were applied for evaluating the human exposure or source apportionment problems. The air quality model support center website (http://aqmc.epa.gov.tw) is installed and maintained in accord with TEPA funding support. The website collected several used and available models to lead researchers to estimate the air quality or personal exposure. Some models were released from US Environmental Protection Agency (USEPA), yet some models were developed or improved by Taiwan researchers. Industrial source complex short-term dispersion model version 3 (ISCST3) is widely used for estimating the grid-based exposures (Chio et al. 2014b). AERMOD model developed by American Meteorological Society (AMS) and EPA Regulatory model Improvement Committee (AERMIC) is also a dispersion-based model with more improvements compared with the ISCST3 model. Gaussian trajectory transfer-coefficient modeling systems (GTx) can be used to model the pollutant levels and associated trajectories with forward and backward approaches (Tsuang 2003). The model results can also be applied to quantify the risks of mortality (Kuo et al. 2014). Trajectory photochemical air quality mode (TPAQM) can model ozone production with VOCs and NOx precursors. Taiwan air quality model (TAQM) is based on USEPA's regional acid deposition model (RADM).

On the other hand, many receptor-based models were applied for PM source apportionment. The approaches include Bayesian model, chemical mass balance (CMB), principal component analysis (PCA), positive matrix factorization (PMF), and others (Chan et al. 1996; Chio et al 2004; Liao et al 2013). Furthermore, Chio et al. (2007, 2014a) integrated the PM source contributions into risk assessment issues. Lee et al. (2014) applied the land-use regression (LUR) models to estimate personal SOx and SOx exposures in Taipei. In the near future, several cohorts can be established and combined with LUR models to construct the precise exposure–effect relationships.

2.5 CONCLUSION

The epidemiological, panel, and experimental studies in Taiwan have provided the sufficient evidence that short-term exposures to air pollution, especially PM, are associated with a variety of adverse cardiopulmonary effects. Several studies also demonstrate that not only particulate mass concentration but also chemical constituents contribute to PM-related cardiovascular effects. Individual susceptibility and meteorological factors can modify the effects of air pollution on cardiovascular systems.

Nevertheless, several priority research issues should be addressed in future studies. One important issue is the health effect of long-term air pollution exposures. Only one study with secondary data analysis demonstrated that increased 1-year-averaged PM_{10}, $PM_{2.5}$, O_3, and NO_2 were associated with elevated systolic and diastolic BP (Chuang et al., 2011). However, using nearby AQM station to estimate personal long-term air pollution exposures probably resulted in measurement error. The research team from National Taiwan University participated in the European Study of Cohorts for Air Pollution Effects (ESCAPE) project—the most up-to-date, largest multicenter study in the European union—and developed the LUR model to estimate residential long-term exposures to PM and NO_x (Lee et al., 2014). Some important findings in long-term exposures to air pollution on cardiovascular system have been obtained. Using LUR to estimate individual 1-year exposures of PM and NO_x, the results of a population-based study demonstrate that 1-year exposures to fine and coarse particles and NO_x are associated with higher diastolic BP. And such associations are stronger among susceptible subjects who are hypertensive, diabetic, or obese (Chen et al., 2014c). Another panel study shows that 1-year exposures to $PM_{2.5}$ absorbance are associated with increase in carotid intima–media thickness, which is an indicator of atherosclerosis (Su et al., 2014).

To date, a large body of evidence is suggestive of a causal relationship between air pollution and a variety of health outcomes; however, other emerging environmental stressors, such as noise, green spaces, and heat islands, were rarely surveyed. Also, individual or neighborhood socioeconomic status and lifestyle are not investigated with sufficient rigor. Such comprehensive assessments require large longitudinal studies and interdisciplinary expertise to harmonize and analyze data spanning from birth to old age.

REFERENCES

Anderson, J. O., Thundiyil, J. G., and Stolbach, A. (2012). Clearing the air: A review of the effects of particulate matter air pollution on human health. *J Med Toxicol*, 8, 166–75.

Brook, R. D., Rajagopalan, S., Pope, C. A. 3rd, Brook, J. R., Bhatnagar, A., Diez-Roux, A. V., Holguin, F., Hong, Y., Luepker, R. V., Mittleman, M. A., Peters, A., Siscovick, D., Smith, S. C. Jr., Whitsel, L., and Kaufman, J. D. (2010). American Heart Association Council on Epidemiology and Prevention, Council on the Kidney in Cardiovascular Disease, and Council on Nutrition, Physical Activity and Metabolism. Particulate matter air pollution and cardiovascular disease: An update to the scientific statement from the American Heart Association. *Circulation*, 121, 2331–78.

Carbon Monitoring for Action. (2014). Taichung. Available online http://carma.org/plant.

Carlsten, C. and Georas, S. N. (2014). Update in environmental and occupational lung diseases 2013. *Am J Respir Crit Care Med*, 189, 1037–43.

Chan, C. C., Chuang, K. J., Chen, W. J., Chang, W. T., Lee, C. T., and Peng, C. M. (2008). Increasing cardiopulmonary emergency visits by long-range transported Asian dust storms in Taiwan. *Environ Res*, 106, 393–400.

Chan, C. C., Chuang, K. J., Chien, L. C., Chen, W. J., and Chang, W. T. (2006). Urban air pollution and emergency admissions for cerebrovascular diseases in Taipei, Taiwan. *Eur Heart J*, 27, 1238–44.

Chan, C. C., Chuang, K. J., Shiao, G. M., and Lin, L. Y. (2004). Personal exposure to submicrometer particles and heart rate variability in human subjects. *Environ Health Perspect*, 112, 1063–7.

Chan, C. C. and Ng, H. C. (2011). A case-crossover analysis of Asian dust storms and mortality in the downwind areas using 14-year data in Taipei. *Sci Total Environ*, 410–11, 47–52.

Chan, C. C., Nien, C. K., and Hwang, J. S. (1996). Receptor modeling of VOCs, CO, NOx, and Thc in Taipei. *Atmos Environ*, 30, 25–33.

Chan, C. C. and Wu, T. H. (2005). Effects of ambient ozone exposure on mail carriers' peak expiratory flow rates. *Environ Health Perspect*, 113, 735–8.

Chang, C. C., Hwang, J. S., Chan, C. C., Wanf, P. Y., Hu, T. H., and Cheng, T. J. (2004). Effects of concentrated ambient particles on heart rate, blood pressure, and cardiac contractility in spontaneously hypertensive rats. *Inhal Toxicol*, 16, 421–9.

Chang, C. C., Kuo, C. C., Liou, S. H., and Yang, C. Y. (2013). Fine particulate air pollution and hospital admissions for myocardial infarction in a subtropical city: Taipei, Taiwan. *J Toxicol Environ Health A*, 76, 440–8.

Chang, C. C., Tsai, S. S., Ho, S. C., and Yang, C. Y. (2005). Air pollution and hospital admissions for cardiovascular disease in Taipei, Taiwan. *Environ Res*, 98, 114–9.

Chang, Y. K., Wu, C. C., Lee, L. T., Lin, R. S., Yu, Y. H., and Chen, Y. C. (2012). The short-term effects of air pollution on adolescent lung function in Taiwan. *Chemosphere*, 87, 26–30.

Chao, H. J., Chan, C. C., Rao, C. Y., Lee, C. T., Chuang, Y. C., Chiu, Y. H., Hsu, H. H., and Wu, Y. H. (2012). The effects of transported Asian dust on the composition and concentration of ambient fungi in Taiwan. *Int J Biometeorol*, 56, 211–9.

Chen, P. C., Lai, Y. M., Chan, C. C., Hwang, J. S., Yang, C. Y., and Wang, J. D. (1999). Short-term effect of ozone on the pulmonary function of children in primary school. *Environ Health Perspect*, 107, 921–5.

Chen, P. C., Lai, Y. M., Wang, J. D., Yang, C. Y., Hwang, J. S., Kuo, H. W., Huang, S. L., and Chan, C. C. (1998). Adverse effect of air pollution on respiratory health of primary school children in Taiwan. *Environ Health Perspect*, 106, 331–5.

Chen, S. Y., Chan, C. C., Lin, Y. L., Hwang, J. S., and Su, T. C. (2014a). Fine particulate matter results in hemodynamic changes in subjects with blunted nocturnal blood pressure dipping. *Environ Res*, 131, 1–5.

Chen, S. Y., Lin, Y. L., Chang, W. T., Lee, C. T., and Chan, C. C. (2014b). Increasing emergency room visits for stroke by elevated levels of fine particulate constituents. *Sci Total Environ*, 473–74, 446–50.

Chen, S. Y., Su, T. C., Lin, Y. L., and Chan, C. C. (2012). Short-term effects of air pollution on pulse pressure among nonsmoking adults. *Epidemiology*, 23, 341–8.

Chen, S. Y., Wu, C. F., Lee, J. H., Hoffmann, B., Peters, A., Brunekreef, B., Chu, D. C., and Chan, C. C. (2014c). Associations between long-term particulate matter and nitrogen oxides exposures and blood pressure in the elderly population. Submitted to *Environ Health Perspect*.

Chen, Y. M., Lin, W. Y., and Chan, C. C. (2014d). The impact of petrochemical industrialisation on life expectancy and per capita income in Taiwan: An 11-year longitudinal study. *BMC Public Health*, 14, 247.

Chen, Y. S., Sheen, P. C., Chen, E. R., Liu, Y. K., Wu, T.N., and Yang, C. Y. (2004). Effects of Asian dust storm events on daily mortality in Taipei, Taiwan. *Environ Res*, 95, 151–5.

Cheng, M. F., Tsai, S. S., Chiu, H. F., Sung, F. C., Wu, T. N., and Yang, C. Y. (2009a). Air pollution and hospital admissions for pneumonia: Are there potentially sensitive groups? *Inhal Toxicol*, 21, 1092–8.

Cheng, M. F., Tsai, S. S., Wu, T. N., Chen, P. S., and Yang, C. Y. (2007). Air pollution and hospital admissions for pneumonia in a tropical city: Kaohsiung, Taiwan. *J Toxicol Environ Health A*, 70, 2021–6.

Cheng, M. F., Tsai, S. S., and Yang, C. Y. (2009b). Air pollution and hospital admissions for myocardial infarction in a tropical city: Kaohsiung, Taiwan. *J Toxicol Environ Health A*, 72, 1135–40.

Cheng, M. T., Lin, Y. C., Chio, C. P., Wang, C. F., and Kuo, C. Y. (2005). Characteristics of aerosols collected in central Taiwan during an Asian dust event in spring 2000. *Chemosphere*, 61, 1439–50.

Cheng, T. J., Huang, Y. F., and Ma, Y. C. (2001). Urinary thiodiglycolic acid levels for vinyl chloride monomer-exposed polyvinyl chloride workers. *J Occup Environ Med*, 43, 934–8.

Cheng, T. J., Hwang, J. S., Wang, P. Y., Tsai, C. F., Chen, C. Y., Lin, S. H., and Chan, C. C. (2003). Effects of concentrated ambient particles on heart rate and blood pressure in pulmonary hypertensive rats. *Environ Health Perspect*, 111, 147–50.

Chio, C. P., Chen, S. C., Chiang, K. C., Chou, W. C., and Liao, C. M. (2007). Oxidative stress risk analysis for exposure to diesel exhaust particle-induced reactive oxygen species. *Sci Total Environ*, 387, 113–27.

Chio, C. P., Cheng, M. T., and Wang, C. F. (2004). Source apportionment to PM_{10} in different air quality conditions for Taichung urban and coastal areas, Taiwan. *Atmos Environ*, 38, 6893–905.

Chio, C. P., Liao, C. M., Tsai, Y. I., Cheng, M. T., and Chou, W. C. (2014a). Health risk assessment for residents exposed to atmospheric diesel exhaust particles in southern region of Taiwan. *Atmos Environ*, 85, 64–72.

Chio, C. P., Yuan, T. H., Shie, R. H., and Chan, C. C. (2014b). Assessing vanadium and arsenic exposures for residents near a petrochemical complex by two-stage dispersion models. *J Hazard Mater*, 271, 98–107.

Chiu, H. F., Cheng, M. H., and Yang, C. Y. (2009). Air pollution and hospital admissions for pneumonia in a subtropical city: Taipei, Taiwan. *Inhal Toxicol*, 21, 32–7.

Chiu, H. F., Tiao, M. M., Ho, S. C., Kuo, H. W., Wu, T. N., and Yang, C. Y. (2008). Effects of Asian dust storm events on hospital admissions for chronic obstructive pulmonary disease in Taipei, Taiwan. *Inhal Toxicol*, 20, 777–81.

Chiu, H. F., Tsai, S. S., Weng, H. H., and Yang, C. Y. (2013). Short-term effects of fine particulate air pollution on emergency room visits for cardiac arrhythmias: A case-crossover study in Taipei. *J Toxicol Environ Health A*, 76, 614–23.

Chiu, H. F. and Yang, C. Y. (2009). Air pollution and emergency room visits for arrhythmias: Are there potentially sensitive groups? *J Toxicol Environ Health A*, 72, 817–23.

Chiu, H. F. and Yang, C. Y. (2013). Short-term effects of fine particulate air pollution on ischemic stroke occurrence: A case-crossover study. *J Toxicol Environ Health A*, 76, 1188–97.

Chuang, K. J., Chan, C. C., Chen, N. T., Su, T. C., and Lin, L. Y. (2005). Effects of particle size fractions on reducing heart rate variability in cardiac and hypertensive patients. *Environ Health Perspect*, 113, 1693–7.

Chuang, K. J., Chan, C. C., Su, T. C., Lee, C. T., and Tang, C. S. (2007). The effect of urban air pollution on inflammation, oxidative stress, coagulation, and autonomic dysfunction in young adults. *Am J Respir Crit Care Med*, 176, 370–6.

Chuang, K. J., Yan, Y. H., and Cheng, T. J. (2010). Effect of air pollution on blood pressure, blood lipids, and blood sugar: A population-based approach. *J Occup Environ Med*, 52, 258–62.

Chuang, K. J., Yan, Y. H., Chiu, S. Y., and Cheng, T. J. (2011). Long-term air pollution exposure and risk factors for cardiovascular diseases among the elderly in Taiwan. *Occup Environ Med*, 68, 64–8.

CTCI. (2009). *Update and Management of National Air Emission Data System and Establishment of Spatial Distribution in Query*. Environmental Protection Administration, Taiwan. Report No. Epa-97-FA11-03-A176. (*in Chinese*)

Formosa Plastic Group. (2010). *The Environmental Impact of the NG-6 Formosa Petrochemical Corporation: A Ten Year Evaluation*. In Taiwan, Yunlin: Formosa Plastic Group, 2010. (*in Chinese*)

FPCC. (2012). *No. 6 Naphtha Cracking Project – Magnitude and Facilities, Formosa Petrochemical Corporation Taiwan*. (*in Chinese*)

Guo, Y. L., Lin, Y. C., Sung, F. C., Huang, S. L., Ko, Y. C., Lai, J. S., Su, H. J., Shaw, C. K., Lin, R. S., and Dockery, D. W. (1999). Climate, traffic-related air pollutants, and asthma prevalence in middle-school children in Taiwan. *Environ Health Perspect*, 107, 1001–6.

Hsieh, Y. L., Tsai, S. S., and Yang, C. Y. (2013). Fine particulate air pollution and hospital admissions for congestive heart failure: A case-crossover study in Taipei. *Inhal Toxicol*, 25, 455–60.

Hsieh, Y. L., Yang, Y. H., Wu, T. N., and Yang, C. Y. (2010). Air pollution and hospital admissions for myocardial infarction in a subtropical city: Taipei, Taiwan. *J Toxicol Environ Health A*, 73, 757–65.

Hu, S. W., Chan, Y. J., Hsu, H. T., Wu, K. Y., Changchien, G. P., Shie, R. H., and Chan, C. C. (2011). Urinary levels of 1-hydroxypyrene in children residing near a coal-fired power plant. *Environ Res*, 111, 1185–91.

Huang, C. H., Lin, L. Y., Tsai, M. S., Hsu, C. Y,. Chen, H. W., Wang, T. D., Chang, W. T., Cheng, T. J., and Chen, W. J. (2010). Acute cardiac dysfunction after short-term diesel exhaust particles exposure. *Toxicol Let*, 192, 349–55.

Hwang, B. F., Jaakkola, J. J., Lee, Y. L., Lin, Y. C., and Guo, Y. L. (2006). Relation between air pollution and allergic rhinitis in Taiwanese schoolchildren. *Respir Res*, 7, 23.

Kang, J. H., Liu, T. C., Keller, J., and Lin, H. C. (2013). Asian dust storm events are associated with an acute increase in stroke hospitalisation. *J Epidemiol Community Health*, 67, 125–31.

Kuo, P. H., Tsuang, B. J., Chen, C. J., Hu, S. W., Chiang, C. J., Tsai, J. L., Tang, M. L., Chen, G. I., and Ku, K. C. (2014). Risk assessment of mortality for all-cause, ischemic heart disease, cardiopulmonary disease, and lung cancer due to the operation of the world's largest coal-fired power plant. *Atmos Environ*, 96, 117–24.

Lee, C. T., Chuang, M. T., Chan, C. C., Cheng, T. J., and Huang, S. L. (2006). Aerosol characteristics from the Taiwan aerosol supersite in the Asian yellow-dust periods of 2002. *Atmos Environ*, 40, 3409–18.

Lee, I. M., Tsai, S. S., Chang, C. C., Ho, C. K., and Yang, C. Y. (2007a). Air pollution and hospital admissions for chronic obstructive pulmonary disease in a tropical city: Kaohsiung, Taiwan. *Inhal Toxicol*, 19, 393–8.

Lee, I. M., Tsai, S. S., Ho, C. K., Chiu, H. F., and Yang, C. Y. (2007b). Air pollution and hospital admissions for congestive heart failure in a tropical city: Kaohsiung, Taiwan. *Inhal Toxicol*, 19, 899–904.

Lee, J. H., Wu, C. F., Hoek, G., DE Hoogh, K., Beelen, R., Brunekreef, B., and Chan, C. C. (2014). Land use regression models for estimating individual NO_x and NO_2 exposures in a metropolis with a high density of traffic roads and population. *Sci Total Environ*, 472, 1163–71.

Lee, M. W., Chen, M. L., Lung, S. C. C., Tsai, C. J., Yin, X. J., and Mao, I. F. (2010). Exposure assessment of $PM_{2.5}$ and urinary 8-OHdG for diesel exhaust emission inspector. *Sci Total Environ*, 408, 505–10.

Lee, Y. L., Shaw, C. K., Su, H. J., Lai, J. S., Ko, Y. C., Huang, S. L., Sung, F. C., and Guo, Y. L. (2003). Climate, traffic-related air pollutants and allergic rhinitis prevalence in middle-school children in Taiwan. *Eur Respir J*, 21, 964–70.

Lee, Y. L., Wang, W. H., Lu, C. W., Lin, Y. H., and Hwang, B. F. (2011). Effects of ambient air pollution on pulmonary function among schoolchildren. *Int J Hyg Environ Health*, 214, 369–75.

Lei, Y. C., Chan, C. C., Wang, P. Y., Lee, C. T., and Cheng, T. J. (2004). Effects of Asian dust event particles on inflammation markers in peripheral blood and bronchoalveolar lavage in pulmonary hypertensive rats. *Environ Res*, 95, 71–6.

Liang, W. M., Wei, H. Y., and Kuo, H. W. (2009). Association between daily mortality from respiratory and cardiovascular diseases and air pollution in Taiwan. *Environ Res*, 109, 51–8.

Liao, H. T., Kuo, C. P., Hopke, P. K., and Wu, C. F. (2013). Evaluation of a modified receptor model for solving multiple time resolution equations: A simulation study. *Aerosol Air Qual Res*, 13, 1253–62.

Lin, R. S., Sung, F. C., Huang, S. L., Gou, Y. L., Ko, Y. C., Gou, H. W., and Shaw, C. K. (2001). Role of urbanization and air pollution in adolescent asthma: A mass screening in Taiwan. *J Formos Med Assoc*, 100, 649–55.

Liu, S. C. and Shiu, C. J. (2001). Asian dust storms and their impact on the air quality of Taiwan. *Aerosol Air Qual Res*, 1, 1–8.

Pan, C. H., Chan, C. C., Huang, Y. L., and Wu, K. Y. (2008). Urinary 1-hydroxypyrene and malondialdehyde in male workers in Chinese restaurants. *Occup Environ Med*, 65, 732–5.

Pan, H. H., Chen, C. T., Sun, H. L., Ku, M. S., Liao, P. F., Lu, K. H., Sheu, J. N., Huang, J. Y., Pai, J. Y., and Lue, K. H. (2014). Comparison of the effects of air pollution on outpatient and inpatient visits for asthma: A population-based study in Taiwan. *PLoS One*, 9, e96190.

Shie, R. H. and Chan, C. C. (2013). Tracking hazardous air pollutants from a refinery fire by applying on-line and off-line air monitoring and back trajectory modeling. *J Hazard Mater*, 261, 72–82.

Shie, R. H., Yuan, T. H., and Chan, C. C. (2013). Using pollution roses to assess sulfur dioxide impacts in a township downwind of a petrochemical complex. *J Air Waste Manage Assoc*, 63, 702–11.

Su, T. C., Chan, C. C., Liau, C. S., Lin, L. Y., Kao, H. L., and Chuang, K. J. (2006). Urban air pollution increases plasma fibrinogen and plasminogen activator inhibitor-1 levels in susceptible patients. *Eur J Cardiovasc Prev Rehabil*, 13, 849–52.

Su, T. C., Hwang, J. J., Shen, Y. C., and Chan, C. C. (2014). Carotid intima-media thickness is associated with long-term exposure 1 to traffic-2 related air Pollution in Middle-aged adults. *Environ Health Perspect* (Accepted).

Taiwan Power Company. (2011). *The World-Largest Coal-Fired Power Plant: Taichung Power Plant, Yuan Magazine*. http://dept.taipower.com.tw/yuan/89/P4.pdf

Tang, C. S., Chang, L. T., Lee, H. C., and Chan, C. C. (2007). Effects of personal particulate matter on peak expiratory flow rate of asthmatic children. *Sci Total Environ*, 382, 43–51.

Tsai, D. H., Wang, J. L., Chuang, K. J., and Chan, C. C. (2010). Traffic-related air pollution and cardiovascular mortality in central Taiwan. *Sci Total Environ*, 408, 1818–23.

Tsai, S. S., Chang, C. C., Liou, S. H., and Yang, C. Y. (2014a). The effects of fine particulate air pollution on daily mortality: A case-crossover study in a subtropical city, Taipei, Taiwan. *Int J Environ Res Public Health*, 11, 5081–93.

Tsai, S. S., Chang, C. C., and Yang, C. Y. (2013). Fine particulate air pollution and hospital admissions for chronic obstructive pulmonary disease: A case-crossover study in Taipei. *Int J Environ Res Public Health*, 10, 6015–26.

Tsai, S. S., Chen, C. C., and Yang, C. Y. (2014b). Short-term effect of fine particulate air pollution on daily mortality: A case-crossover study in a tropical city, Kaohsiung, Taiwan. *J Toxicol Environ Health A*, 77, 467–77.

Tsai, S. S., Cheng, M. H., Chiu, H. F., Wu, T. N., and Yang, C. Y. (2006). Air pollution and hospital admissions for asthma in a tropical city: Kaohsiung, Taiwan. *Inhal Toxicol*, 18, 549–54.

Tsai, S. S., Chiu, H. F., Wu, T. N., and Yang, C. Y. (2009). Air pollution and emergency room visits for cardiac arrhythmia in a subtropical city: Taipei, Taiwan. *Inhal Toxicol*, 21, 1113–8.

Tsai, S. S., Goggins, W. B., Chiu, H. F., and Yang, C. Y. (2003a). Evidence for an association between air pollution and daily stroke admissions in Kaohsiung, Taiwan. *Stroke*, 34, 2612–6.

Tsai, S. S., Huang, C. H., Goggins, W. B., Wu, T. N., and Yang, C. Y. (2003b). Relationship between air pollution and daily mortality in a tropical city: Kaohsiung, Taiwan. *J Toxicol Environ Health A*, 66, 1341–9.

Tsai, S. S. and Yang, C. Y. (2014). Fine particulate air pollution and hospital admissions for pneumonia in a subtropical city: Taipei, Taiwan. *J Toxicol Environ Health A*, 77, 192–201.

Tsuang, B. J. (2003). Quantification on the source receptor relationship of primary pollutants and secondary aerosols by a Guassion plume trajectory model: Part 1—Theory. *Atmos Environ*, 37, 3981–91.

Wang, K. Y. and Chau, T. T. (2013). An association between air pollution and daily outpatient visits for respiratory disease in a heavy industry area. *PLoS One*, 8, e75220.

Wang, T. N., Ko, Y. C., Chao, Y. Y., Huang, C. C., and Lin, R. S. (1999). Association between indoor and outdoor air pollution and adolescent asthma from 1995 to 1996 in Taiwan. *Environ Res*, 81, 239–47.

Wu, C. F., Kuo, I. C., Su, T. C., Li, Y. R., Lin, L. Y., Chan, C. C., and Hsu, S. C. (2010). Effects of personal exposure to particulate matter and ozone on arterial stiffness and heart rate variability in healthy adults. *Am J of Epidemiol*, 171, 1299–309.

Yan, Y. H., Huang, C. H., Chen, W. J., Wu, M. F., and Cheng, T. J. (2008). Effects of diesel exhaust particles on left ventricular function in isoproterenol-induced myocardial injury and healthy rats. *Inhal Toxicol*, 20, 199–203.

Yang, C. Y. (2008). Air pollution and hospital admissions for congestive heart failure in a subtropical city: Taipei, Taiwan. *J Toxicol Environ Health A*, 71, 1085–90.

Yang, C. Y., Chen, C. C., Chen, C. Y., and Kuo, H. W. (2007). Air pollution and hospital admissions for asthma in a subtropical city: Taipei, Taiwan. *J Toxicol Environ Health A*, 70, 111–7.

Yang, C. Y. and Chen, C. J. (2007). Air pollution and hospital admissions for chronic obstructive pulmonary disease in a subtropical city: Taipei, Taiwan. *J Toxicol Environ Health A*, 70, 1214–9.

Yang, C. Y., Chen, Y. S., Chiu, H. F., and Goggins, W. B. (2005). Effects of Asian dust storm events on daily stroke admissions in Taipei, Taiwan. *Environ Res*, 99, 79–84.

Yang, C. Y., Chen, Y. S., Yang, C. H., and Ho, S. C. (2004). Relationship between ambient air pollution and hospital admissions for cardiovascular diseases in Kaohsiung, Taiwan. *J Toxicol Environ Health A*, 67, 483–93.

Yu, H. L., Yang, C. H., and Chien, L. C. (2013). Spatial vulnerability under extreme events: A case of Asian dust storm's effects on children's respiratory health. *Environ Int*, 54, 35–44.

Yu, J. H., Lue, K. H., Lu, K. H., Sun, H. L., Lin, Y. H., and Chou, M. C. (2005). The relationship of air pollution to the prevalence of allergic diseases in Taichung and Chu-Shan in 2002. *J Microbiol Immunol Infect*, 38, 123–6.

Yuan, T. H., Chio, C. P., Shie, R. H., Pien, W. J., and Chan, C. C. (2014). The distance-to-source trend in vanadium and arsenic exposure for residents living near a petrochemical complex. *J Expo Sci Environ Epidemiol*. (In press)

3 Air Pollution and Health in Japan

Toru Takebayashi

CONTENTS

3.1 BACKGROUND

Japan is located in East Asia, comprising the four main islands and more than 6000 smaller islands with land space of 377, 960 km^2. The total population size is approximately 127 million, and the proportion of elderly population to total population is 25%. Concentration of the population in urban areas is still ongoing, and more than half of the population lives within three major urban areas. For land use, 66.3% is woodland, 12.1% is agricultural, 5.0% is residential, 3.6% is transport, and 3.5% is water (Statistics Japan, 2014).

The Japanese islands extend from north to south as well as east to west over 2000 km and most parts of them are in the temperate zone with four distinct seasons, but the north is cool and the south is subtropical. Climate is different between the Pacific coast side and the Sea of Japan side. Pacific Ocean climatic zone is characterized by hot and humid days during summer because of seasonal south wind from the Pacific Ocean. In the Sea of Japan climatic zone, cold monsoon in winter from the Siberian anticyclone, that is, westerlies, brings heavy snowfalls while relatively dry during summer. Hokkaido climatic zone is also cold in winter and cool in summer. In addition, there are three other climatic zones; Central Highland, Seto Inland Sea, and Ryukyu Islands zones. Annual average (minimum to maximum) daily temperatures of major cities are 9.2 (5.8–13.0) degrees Celsius in Sapporo, 17.1 (13.6–21.0) in Tokyo, 17.1 (13.4–21.5) in Osaka, and 17.7 (14.3–21.8) in Fukuoka, respectively.

Going back to the world history, modernization had brought a good and a bad side for society. Industrialization of Japan started in 1890s, and rapid economic growth has been achieved around 1950–1970s. During that time, the life expectancy prolonged surprisingly: life expectancy for males improved from 58.0 in 1950 to 71.7

in 1970 and that for females did change from 61.5 to 76.9. On the other hand, breakneck industrialization and urbanization caused severe air pollution episodes in Japan. Around 1955–1965, air pollution episodes with respiratory problems were experienced among industrial area residents due to inhalation of soot dust and sulfur dioxide (SO_2) emitted from coal-burning plants and oil-refining plants. The most well-known SO_2 episode in Japan was occurred in Yokkaichi of Mie prefecture, which is located in the Pacific coast side of the central Japan. Numerous petrochemical plants started their operation in 1960, and residents in some areas started to complain of asthma-like symptoms just in a few years. Annual emission of sulfur oxides was estimated to be 130–140 thousand tons in 1963–1964. The Yokkaichi Air Pollution Lawsuit had been filed in 1967 regarding the legal causal relationship between air pollution mainly due to SO_2 and the infringement of the right of health as well as the joint responsibility of six companies in the area, and the decision was made to find joint responsibility of six companies based on epidemiological causality (Yoshida et al., 2007).

To control such situations legally, the Act Concerning Control of Soot and Smoke Emission was enacted in 1962. For basic national policy against environmental pollution, the Basic Law for Environmental Pollution Control was enacted in 1967, leading to establishment of the Air Pollution Control Act in 1968, aiming at protecting the nation's health and to conserve the human environment. Basic framework for legal pollution control was shaped, including setting the environmental quality standard (EQS), monitoring environmental quality, implementing countermeasures for hazardous pollutants, and stipulating allowable limits of emission. In Yokkaichi, introduction and implementation of total emission control for SO_2 had largely contributed to reducing pollution level in the region. Environmental air quality standard for SO_2 was first set as not exceeding annual average for hourly values of 0.05 ppm in 1969, and revised to (1) daily average for hourly values of 0.04 ppm, and (2) hourly value of 0.1 ppm. The Basic Law for Environmental Pollution Control was revised to the Basic Environment Law in 1993, in which formation of the environmental conservation society and the global environmental protection were added as basic policy (Ministry of the Environment [MOE], 2014).

In 1970s, accompanying progress in urbanization, pollution with photochemical oxidants, that is, urban ozone, has been recognized as next air pollution issue, followed by nitrogen oxides (NO_x). Regulations on motor vehicle exhausts were stipulated in the Air Pollution Control Act in 1978, and the Automobile NOx Act was enacted in 1992.

This chapter describes current status and issues of air pollution control in Japan, mainly focusing on particulate matter.

3.2 HEALTH EFFECTS

Three Japanese epidemiologic studies on health effects of long-term exposure to particulates have been published.

3.2.1 Three-Prefecture Cohort Study

The aim of this study (Katanoda et al., 2011) was to investigate health effects of air pollutants on mortality, especially focusing on respiratory diseases. Target population was 100,629 adults over 40 years old (male 46%) in six regions of three prefectures (Miyagi, Aichi, and Osaka). The baseline survey was conducted in 1983–1985, and the participants had been followed for 10–15 years. In this study, concentration of suspended particulate matter (SPM), SO_2, and nitrogen dioxide (NO_2) was continuously measured. In the statistical analysis, particulate matter with an aerodynamic diameter of 2.5 μm ($PM_{2.5}$) concentration was estimated by multiplying SPM concentrations by a factor of 0.7. Average concentrations prior to the study (1974–1983) were 16.8–41.9 μg/m^3 for $PM_{2.5}$, 2.4–19.0 ppb for SO_2, and 1.2–33.7 ppb for NO_2, respectively. The major finding of the study was that lung cancer mortality was significantly associated with SO_2, NO_2, and SPM concentration, adjusting for major potential confounders, including smoking. The hazard ratios (95% confidence interval [CI]) per 10-unit increase in $PM_{2.5}$, SO_2, and NO_2 were 1.24 (1.12–1.37), 1.26 (1.07–1.48), and 1.17 (1.10–1.26), respectively. Other respiratory diseases were also associated with SO_2 and NO_2 in females, but no association was observed with SPM.

3.2.2 NIPPON DATA 80

National Integrated Project for Prospective Observation of Noncommunicable Disease and Its Trends in the Aged 80 (NIPPON DATA 80) is a prospective cohort study of the participants of the National Survey on Circulatory Disorders in 1980 (Ueda et al., 2012). The study population was the participants of the National Nutritional Survey of Japan, which has been performed using weighing record method for three consecutive days to each household in a large-scale sample of representative Japanese from 300 randomly selected districts in Japan. In this report, exposure information was obtained from the nearest monitoring station to each census area using geographic information system (GIS). Averaged particulate concentrations (defined as particles with a 100% cutoff level of aerodynamic diameter 10 μm, which can be considered to be particles of less than 7 μm with a 50% cutoff level) measured at the corresponding station were assigned to each participants residing in the census area. Study subjects were 7250 residents aged 30 years or more, randomly selected 232 census areas in 1980, and were followed up until 2004. The Cox proportional hazard model was applied adjusting for sex, age, body mass index, blood pressure, total cholesterol, blood glucose, smoking categories, drinking categories, and municipality population size. As a result, numbers of deaths were 1716 for all-cause deaths, 571 for cardiovascular deaths, 116 for coronary deaths, and 250 for stroke deaths, but no significant association was observed between long-term exposure to particulates and cardiovascular mortality risk.

3.2.3 Japan Public Health Center–Based Study

The Japan Public Health Center (JPHC)-based study (Nishiwaki et al., 2013) is a prospective cohort study lead by the National Cancer Research Center of Japan in collaboration with 11 JPHCs nationwide. Cohort I started in 1990 with five centers, and Cohort II initiated in 1993–1994 in six centers. Among more than 100,000 participants, 78,057 participants (37,121 males, 40,936 females) aged 40–59 in nine areas where ambient air monitoring stations were equipped near to the public health center were included in the mortality analysis. For incidence analysis, 62,142 (30,238 males, 31,904 females) from seven areas were subject to analysis. Average of annual particulate concentrations (defined as the same as NIPPION Data80 study) during the study period were ranged 17.2–43.7 μg/m^3. After controlling for major potential confounders including age, sex, smoking, body mass index, alcohol consumption, and blood pressure, no significant association with particulate level was observed in mortality analysis, but risks having coronary heart disease and myocardial infarction were increased in incidence analysis, particularly among smokers. Adjusted hazard ratios per 10 μg/m^3 increase in particulates were 1.39 (95% CI: 1.01–1.93) for coronary heart disease and 1.52 (1.08–2.13) for myocardial infarction among smokers. Similar results were obtained for females, although the associations diminished when one center area was excluded from the analysis.

There exist methodological issues to examine long-term effects of particulate matter on cardiovascular diseases. First, distribution pattern of coronary heart disease and stroke is contradictory between Japan and Western countries because of different distribution patterns of major risk factors such as serum cholesterol and prevalence of obesity. The leading cause of cardiovascular deaths is stroke in Japan, while it is coronary heart disease in Westerns. Second, different spatial distribution patterns of cardiovascular mortality and air particulate matter concentration between urban and rural areas makes it more difficult to conduct a valid epidemiological study in Japan because incidence and mortality of stroke tend to be high in rural areas indicating that general risk factors are more prevalent among rural residents while particulate matter levels are to be high in urban areas in Japan. A valid epidemiological study should be conducted in Japan, in which cardiovascular risk factors are comparable except for particulate matter concentrations.

For short-term effects of particulate matter, Ueda et al. reported the results of time-series and case-crossover analyses in the 20 cities of Japan where $PM_{2.5}$ concentration has been monitored (Ueda et al., 2009; MOE, 2008). Mortality data between January 1, 2002, and December 31, 2004, were obtained through the centrally collected death certificates in the Ministry of Health, Labor, and Welfare in which the cause of death was coded in the National Vital Statistics Bureau. Air quality monitoring data were $PM_{2.5}$ measured by tapered element oscillating microbalance (TEOM), common pollutants, temperature, and humidity. Risk estimates corresponding to $PM_{2.5}$ increase in 10 μg/m^3 were calculated for each city, and then unified risk estimate was obtained with both single-pollutant model and multipollutant model using the generalized additive model. In a single-pollutant model, unified risk ratios for 10 μg/m^3-increase were 1.002 (95% CI: 0.998–1.006) for all-cause death, 0.999 (0.990–1.007) for respiratory death, and 1.001 (0.993–1.009) for cardiovascular death, respectively (Lag = 0). Risk ratio for respiratory death was significantly increased (1.010) at lag 3. Risk ratios of daily mortality for 10 μg/m^3-increase in $PM_{2.5}$ were exceeded 1.0 in some cities, and its unified risk ratio was significantly increased for respiratory deaths. However, no increased risk was observed for cardiovascular deaths. The 98th percentile of 24-hour concentrations of $PM_{2.5}$ ranged 31–55 μg/m^3 for overall analysis, and the

98th $PM_{2.5}$ concentration ranged 44–47 μg/m^3 in the region where daily mortality was significantly increased.

The fatality rate ratios of cerebro-cardiovascular events with relation to daily level of ambient air pollutants were reported using the stroke and acute myocardial infarction (AMI) registry of Takashima county in Shiga prefecture, which is located in the central part of Japan (Turin et al., 2012a). During the study period of 1988–2004, 307 (153 in men and 154 in women) cases had fatal stroke within 28 days of onset among 2038 first-ever stroke cases. The number of fatal AMI cases within 28 days of onset in the same period was 142 (men: 94 and women: 54) among the 429 first-ever AMI events. A Poisson regression model was applied to calculate fatality rate ratios associated with stroke and AMI on a day with higher level of pollutants compared with a day with the lowest level. Daily average pollutant concentration, including SPM, NO_2, SO_2, and Ox (photochemical oxidants), was measured at the nearest monitoring stations, and then all 6210 data of daily average values were divided into quartiles. Multiple-pollutant models with several possible confounders were constructed. Mean values of daily average concentrations were 26.9 μg/m^3 for SPM, 3.9 ppb for SO_2, 16.0 ppb for NO_2, and 28.4 ppb for Ox. Fatality rate ratios for stroke were significantly increased with relation to NO_2 levels while other pollutant levels, including SPM, did not show any association with fatality rate ratios for stroke or AMI.

In addition, time-stratified, bidirectional, case-crossover analysis was applied to the same dataset using the distributed lag model to estimate the effect of pollutant exposure 0–3 days before the onset controlling for meteorological factors (Turin et al., 2012b). The authors concluded that, although an association between SO_2 and hemorrhagic stroke was observed, they found inconclusive evidence for a short-term effect of air pollution on the incidence of other stroke types and AMI in Japan.

Asian dust, transported from the desert areas of China and Mongolia and called kosa in Japan, is also public concern for its possible health effects. In 2014, two studies were reported regarding this issue. A correlational study of subjective symptoms and serum immunoglobulin E (IgE) levels with relation to 3-day Asian dust event in April of 2012 among 25 healthy volunteers was done in Yonago of Tottori prefecture, located in the Sea of Japan side of the western part of Japan (Otani et al., 2014). During the event period, Asian dust-related symptoms (nasal, pharyngeal, ocular, respiratory, and skin) were recorded daily. Serum nonspecific IgE and 33 allergen-specific IgEs were also measured after the dust event. The authors suggested the potential associations between nasal symptoms and fungal allergen IgE level and Asian dusts.

A time-stratified, case-crossover study to examine the association between Asian dust and the incidence of AMI was done in Fukuoka, Kita-Kyusyu, and Kurume cities of northern Kyusyu in western Japan (Matsukawa et al., 2014). A conditional logistic regression analysis was done with 3068 consecutive patients who were hospitalized due to AMI from four different AMI centers between April 2003 and December 2010 to estimate the risk of AMI associated with occurrence of Asian dust event, controlling for ambient temperature and relative humidity. Asian dust event was reported when its visibility reduced to less than 10 km by ground-level observation at the local meteorological observatory. Odds ratio [OR] of AMI incidence was significantly associated with Asian dust event with 4 days of lag (OR 1.33, 95% CI: 1.05–1.69). The association remained significant with cumulative lags of 0–4 days (OR 1.20, 95% CI: 1.02–1.40).

3.3 EXPOSURES, RISK ASSESSMENT, AND RISK MANAGEMENT

The EQS is defined pursuant to the Article 16 of the Basic Environment Law as follows:

1. With regard to the environmental conditions related to air pollution, water pollution, soil contamination, and noise, the government shall respectively establish EQSs, the maintenance of which is desirable for the protection of human health and the conservation of the living environment.
2. In the event that the standards referred to in the preceding paragraph establish more than one category and stipulate that land or water areas to which those categories are to be applied should be designated, the government may delegate to the prefectural governors concerned the authority to designate those land or water areas, in accordance with Cabinet Order.
3. With regard to the standards set forth in point 1, due scientific consideration shall always be given and such standards shall be revised whenever necessary.
4. The government shall make efforts to attain the standard provided in point 1 by comprehensively and effectively implementing policies concerning environmental pollution control which are set forth in this chapter.

MOE set six EQSs for classical pollutants including $PM_{2.5}$ and five EQSs for hazardous air pollutants including dioxin in Japan (Table 3.1). For $PM_{2.5}$, its EQS was set (1) less than or equal to 15 μg/m^3 as annual average and (2) less than or equal to 35 μg/m^3 as daily average (Table 3.2). The expert committee on $PM_{2.5}$ air EQS summarized its discussion as follows:

- There exists the health risk of $PM_{2.5}$, but no threshold was detected in a population level from exposure–response analysis of the existing epidemiological studies, and residual risk may exist even in background regions in Japan, where annual mean $PM_{2.5}$ concentrations ranged 6–12 μg/m^3 although health risk cannot be quantified in such lower level.
- Although discrepancy of cardiovascular effects between United States and Japan can be interpreted by differences in common cardiovascular risk factors or disease pattern, toxicological and epidemiological findings have been accumulated to support cardiovascular effects of $PM_{2.5}$ and it is difficult to fully explain the discrepancy at present, such as by differences in common non-PM risk factors, why we Japanese cannot observe increased cardiovascular risks of PM.
- Size of $PM_{2.5}$ risk is relatively small but the risks should be reduced from the public health point of view, and it is noteworthy that risks of

TABLE 3.1
Air Environmental Quality Standard (EQS) in Japan

Pollutant	Measurement Condition	Standard Value
Sulfur dioxide	Daily average for hourly values	0.04 ppm
	Hourly values	0.1 ppm
Carbon monoxide	Daily average for hourly values	10 ppm
	Average of hourly values for any consecutive 8 hours	20 ppm
Suspended particulate matter	Daily average for hourly values	0.10 mg/m^3
	Hourly values	0.20 mg/m^3
Nitrogen dioxide	Daily average for hourly values	Within 0.04–0.06 ppm zone or below that zone
Photochemical oxidants	Hourly values	0.06 ppm
Fine particulate matter ($PM_{2.5}$)	Annual average	15 μg/m^3
	Daily average for hourly values	35 μg/m^3(the annual 98th percentile values at designated monitoring sites in an area)
Benzene	Annual average	0.003 mg/m^3
Trichloroethylene	Annual average	0.2 mg/m^3
Tetrachloroethylene	Annual average	0.2 mg/m^3
Dichloromethane	Annual average	0.15 mg/m^3
Dioxin	Annual average	0.6 pg-TEQ/m^3

TABLE 3.2
Environmental Quality Standard (EQS) for $PM_{2.5}$ (MOE Notification, September 9, 2009)

I. EQS for $PM_{2.5}$
1. Annual average of less than or equal to 15 μg/m^3
2. Daily average of less than or equal to 35 μg/m^3

II. Evaluation of attainment of $PM_{2.5}$-EQS
1. Annual average: mean of 1-year measurements
2. Daily average: annual 98th percentile of 24-hour concentrations

III. Achievement period
1. The EQS is to be maintained or if possible, achieve at an early date

cardiovascular diseases could be apparent even in Japan in the future if prevalence of cardiovascular risk factors become comparable with that of Western nations.

To maintain air qualities, the Air Pollution Control Act is placed under Basic Environment Law. Protection of the public health and preservation of the human living environment with respect to air pollution are provided by (1) controlling emissions of soot and smoke, volatile organic compounds (VOCs) and particulate by setting emission standards from the business activities of factory and business establishments; (2) promoting various measures concerning hazardous air pollutants; (3) setting allowable limits for automobile exhaust gases; and (4) monitoring of the level of air pollutions. It also provides a liability regime for health damage caused by air pollution from business activities to help victims of air pollution–related health damage.

The emission standards (general emission standard, special emission standard, stricter emission standard, and total mass emission control standard) are set for "soot and smoke" emitted from the business activities of factory and business establishments, including sulfur oxides generated by the combustion of fuel or other items, soot and dust generated by the combustion of fuel or other items or by the use of electricity as a source of heat, cadmium and its compounds, chlorine and hydrogen chloride, fluoride, hydrogen fluoride and silicon fluoride, lead and its compounds, and nitrogen oxides. VOCs such as painting, cleaning, or printing solvents generate photochemical oxidants and particulate matter, and have been subject to regulation since amendment of the Air Pollution Control Law in 2004. Control efforts for VOC emissions have been done through voluntary efforts by business activities with legal restrictions, and significant reduction had been achieved by FY 2010. Emissions of particulates including asbestos from the business activities of factory and business establishments are also regulated under the Act.

The "hazardous air pollutant" is defined in this Act as any substance that is likely to harm human health if ingested continuously and that is a source of air pollution other than soot and smoke or particulates specified in the Act. There are 248 substances present that are being classified as hazardous air pollutants, and 23 of those are classified as priority substances. Those are acrylonitrile, acetaldehyde, arsenic and its inorganic compound, benzene, benzo(a)pyrene, beryllium and its compound,1,3-butadiene, chloroform, chromium(III) and its compound, chromium(VI) compound, 1,2-dichloroethane, dichloromethane, dioxin, ethylene oxide, formaldehyde, manganese and its compound, mercury and its compound, methyl chloride, nickel compound, tetrachloroethylene, toluene, trichloroethylene and vinyl chloride monomer. Among those, air EQS has been set for benzene, dichloromethane, tetrachloroethylene, trichloroethylene, and dioxin, and the emission standards have also been set for benzene, tetrachloroethylene, and trichloroethylene. A special law, Law Concerning Special Measures against Dioxins, was enacted in 1999 for dioxin. The guideline values have been set for nine hazardous air pollutants (Table 3.3). The value is not the standard value (EQS) but the reference value to reduce potential human health risks as it could be set even if scientific evidence is limited to quantify dose–response relationships between exposure to hazardous air pollutant and health.

To control motor vehicle exhaust emissions, various countermeasures are being implemented by MOE as follows: (1) controls on emissions per vehicle; (2) enactment of the Law Concerning Special Measures for Total Emission Reduction of Nitrogen

Oxides and Particulate Matters from Automobiles in Specified Areas and the Act on Regulation, and so on of emissions from nonroad special motor vehicles; and (3) encouraging use of low-pollution vehicles.

Monitoring of air quality has been done through national and local (prefectural) monitoring stations across the country for both classical pollutants and hazardous air pollutants. The number of monitoring stations is more than 1500 for general monitoring and about 400 for roadside monitoring. Recent results (FY 2012) of air quality monitoring of classical pollutants and some hazardous pollutants with EQS or guideline value are shown in Tables 3.4 and 3.5. For photochemical oxidants, its annual mean of the daytime maximal hourly values has continued to increase gradually since 1980s (Figure 3.1), but various countermeasures against VOC and other substances have been taken to lower its daytime concentration.

TABLE 3.3
Guideline Values for Nine Hazardous Air Pollutants

Pollutant	Guideline Value (Annual Mean)
Acrylonitrile	2 μg/m³
Arsenic and its inorganic compound	6 ng-As/m³
1,3-Butadiene	2.5 μg/m³
Chloroform	18 μg/m³
1,2-Dichloroethane	1.6 μg/m³
Manganese and its compound	0.14 μg-Mn/m³
Mercury and its compound	0.04 μg-Hg/m³
Nickel compound	0.025 μg-Ni/m³
Vinyl chloride monomer	10 μg/m³

TABLE 3.4
Results of Air Quality Monitoring of Classical Pollutants in FY 2012

Pollutant	Number of Stations	Number of Stations Exceeding Environmental Quality Standard (EQS)	Overall Average of Annual Mean
NO_2	1285/406	0/3	0.011 ppm/0.031 ppm
SPM	1320/394	4/1	0.019 mg/m³/0.021 mg/m³
Photochemical oxidants	1143/30	1138/30	0.032 ppm/0.027 ppm[a]
SO_2	1022/59	62/0	0.002 ppm/0.002 ppm
CO	68/241	0/0	0.3 ppm/0.4 ppm
$PM_{2.5}$	312/123	177/82	14.5 μg/m³/15.4 μg/m³

[a] Annual mean of daytime maximal hourly values.

TABLE 3.5
Results of Air Quality Monitoring of Some Hazardous Pollutants with Environmental Quality Standard (EQS) or Guideline Value in FY 2012

Pollutant	Number of Stations	Number of Stations Exceeding EQS/ Guideline	Overall Average of Annual Mean
Benzene	419	0	1.2 μg/m^3
Dichloromethane	366	0	1.6 μg/m^3
Tetrachloroethylene	369	0	0.18 μg/m^3
Trichloroethylene	367	0	0.50 μg/m^3
Acrylonitrile	335	0	0.080 μg/m^3
Arsenic compound	280	4	1.5 ng-As/m^3
1,3-Butadiene	374	0	0.14 μg/m^3
Chloroform	334	0	0.20 μg/m^3
1,2-Dichloroethane	347	2	0.17 μg/m^3
Mercury and compound	270	0	2.1 ng-Hg/m^3
Nickel compound	282	0	4.1 ng-Ni/m^3
Vinyl chloride monomer	341	0	0.047 μg/m^3

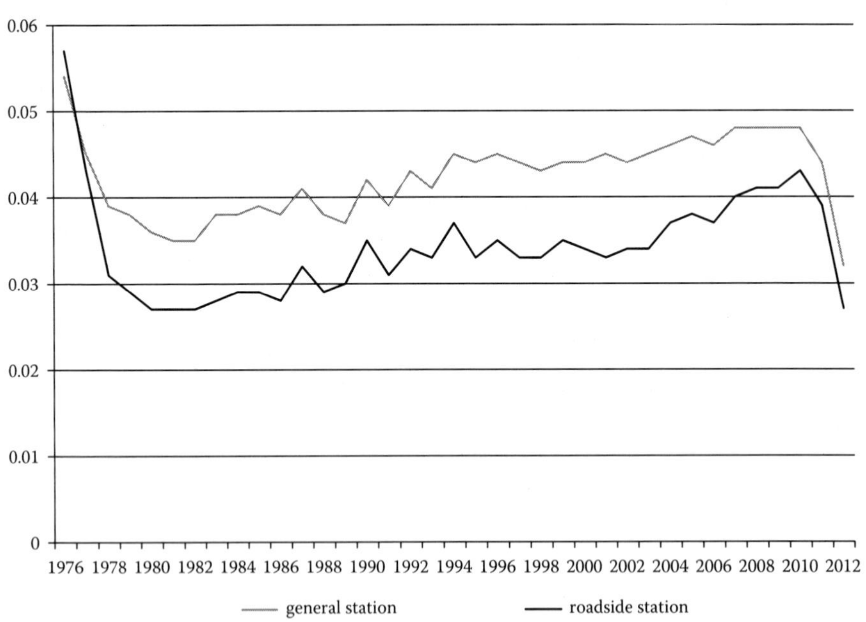

FIGURE 3.1 Trend of annual mean of daytime maximal hourly values for photochemical oxidants concentration in Japan: FY 1976–2012.

3.4 CONCLUSION

Modernization of the society (i.e., industrialization and urbanization) brought various good things for human health such as better access to medical care or improvement of infant and maternal mortality which lead to longevity. However, as Japan had already experienced, uncoordinated rapid industrialization and urbanization resulted in environment pollution and poor air quality. Nowadays, environmental issues are occurring beyond the borders of one country and into neighboring areas. In Japan, $PM_{2.5}$ and photochemical oxidants are the two major remaining air quality issues to be worked on, but we also would like to share our experiences and technologies with the Asian countries to solve air quality issues around this region.

REFERENCES

Katanoda, K., Sobue, T., Satoh, H., Tajima, K., Suzuki, T., Nakatsuka, H., Takezaki, T., Nakayama, T., Nitta, H., Tanabe, K., and Tominaga, S. (2011). An association between long-term exposure to ambient air pollution and mortality from lung cancer and respiratory diseases in Japan. *J Epidemiol*, 21, 132–43.

Matsukawa, R., Michikawa, T., Ueda, K., Nitta, H., Kawasaki, T., Tashiro, H., Mohri, M., and Yamamoto, Y. (2014). Desert dust is a risk factor for the incidence of acute myocardial infarction in Western Japan. *Circ Cardiovasc Qual Outcomes*, 7, 743–8.

Ministry of the Environment. (2008). Report on the Committee on Assessment of the Health Effects of Fine Particulate Matter on Public Health. Japan.

Ministry of the Environment. (2014). *Japan's Regulations and Environmental Law* [Online]. Japan. Available: http://www.env.go.jp/en/coop/pollution.html.

Nishiwaki, Y., Michikawa, T., Takebayashi, T., Nitta, H., Iso, H., Inoue, M., Tsugane, S.; and Japan Public Health Center-Based Prospective Study. (2013). Long-term exposure to particulate matter in relation to mortality and incidence of cardiovascular disease: The JPHC Study. *J Atheroscler Thromb*, 20, 296–309.

Otani, S., Onishi, K., Mu, H., Hosoda, T., Kurozawa, Y., and Ikeguchi, M. (2014). Associations between subjective symptoms and serum immunoglobulin E levels during Asian dust events. *Int J Environ Res Public Health*, 11, 7636–41.

Statistics Japan. (2014). Japan. Available at http://www.stat.go.jp/english/index.htm.

Turin, T. C., Kita, Y., Rumana, N., Nakamura, Y., Ueda, K., Takashima, N., Sugihara, H., Morita, Y., Ichikawa, M., Hirose, K., Nitta, H., Okayama, A., Miura, K., and Ueshima, H. (2012a). Ambient air pollutants and acute case-fatality of cerebro-cardiovascular events: Takashima Stroke and AMI Registry, Japan (1988–2004). *Cerebrovasc Dis*, 34, 130–9.

Turin, T. C., Kita, Y., Rumana, N., Nakamura, Y., Ueda, K., Takashima, N., Sugihara, H., Morita, Y., Ichikawa, M., Hirose, K., Nitta, H., Okayama, A., Miura, K., and Ueshima, H. (2012b). Short-term exposure to air pollution and incidence of stroke and acute myocardial infarction in a Japanese population. *Neuroepidemiology*, 38, 84–92.

Ueda, K., Nagasawa, S. Y., Nitta, H., Miura, K., Ueshima, H., and Group, N. D. R. (2012). Exposure to particulate matter and long-term risk of cardiovascular mortality in Japan: NIPPON DATA80. *J Atheroscler Thromb*, 19, 246–54.

Ueda, K., Nitta, H., and Ono, M. (2009). Effects of fine particulate matter on daily mortality for specific heart diseases in Japan. *Circ J*, 73, 1248–54.

Yoshida, K., Morio, K., and Yokoyama, K. (2007). Epidemiology and environmental pollution: A lesson from Yokkaichi asthma, Japan. In Willis, I. C. (ed.) *Progress in Environmental Research*. Nova Science Publishers, Inc.

4 Air Pollution in Indonesia
Estimates and Surveys

Umar-Fahmi Achmadi and Rachmadhi Purwana

CONTENTS

4.1 BACKGROUND

4.1.1 Demographics

Indonesia is an archipelago of about 17,508 islands laying along the equator in Southeast Asia. Large and small islands are spread throughout the country that will make a jet take 5 to 6 hours for a nonstop flight from one end at the city of Sabang to reach the other far end of the country at the city of Merauke.

Geologically, there are four kinds of islands in the country, namely coral islands, volcanic islands, tectonic islands, and the mix of these three (Achmadi, 2010). This archipelago country is also rich with huge tropical forests and active volcanoes. Therefore from time to time reports of forest fire and volcanic eruption are common in the media and create the problems of air pollution, competing with the existing man-made urban air pollution.

The total population is reported at 253.6 million in 2014 and is projected to increase to more than 274 million in the year 2025 (Bureau of Statistics Indonesia, 2014). According to a report in the year 2000, the total population living in urban areas was 47 million people. At that time, the number was projected to increase to approximately 187 million people in 2025.

From these figures, it is predicted that around 68% of the total Indonesian population will live in cities in 2025, as compared with approximately 23% in the year of 2000. Further, this high concentration of urban new population together with its changes of their social lives and income per capita will become factors contributing to the increase of pollution, including the air pollution in the urban area.

It is reported that 60% of the country's total population live in the island of Java, which accounts for less than 7% of Indonesia's total land area. The island of Java has large cities such as Jakarta, Jogjakarta, Surabaya, Semarang, and Bandung.

Jakarta, the capital of the country has about 9.6 million population. Together with its surrounding municipalities, namely Tangerang, Depok, Bogor, and Bekasi, the total population of this area is more than 23 million people. Except for the Bandung metropolitan city located on top of mountainous area with its population of 2.5 million (Bureau of Statistics Indonesia, 2014), other highly populated cities in Indonesia are located in the coastal area.

Outside the island of Java, there are other large cities like Medan, Makassar, Bandar Lampung, and Palembang, each with their overcrowded population of more than 2 million, reflecting potential air pollution problems.

As an area of sea and ocean with many islands, Indonesia has a tropical pattern of weather with rainfall and wind that to some degree have their impacts on air pollution problems of the country. The objective of this chapter is to generally describe the problem of air pollution on health in Indonesia.

Up to this time, no integrated air quality monitoring network with real-time and regional air quality information is developed in the country. Consequently, assessments reports of the degree of urban and volcanic air pollution are limited (Ministry of Environment, 2012; International Forest Fire News, 2000). Nevertheless, the existing limited air quality monitoring systems and studies are still being used to develop part of the strategy of pollution prevention programs in Indonesia. Insofar, studies on the air pollution impacts on public health rely mostly on sporadic surveys or local case studies in the cities.

Since the 1970s, the government of Indonesia has had a commitment to the World Health Organization (WHO) to provide air quality data for the Global Environmental Monitoring System (GEMS) program. The WHO/UNEP project was intended to monitor the air quality of certain selected cities all over the world. The parameters used in this project are suspended particulate matters (SPMs), sulfur dioxide (SO_2), and nitrogen oxides (NO_x). Results of a decade ago indicate SPM and NO_x were two predominant pollutants. In the large cities of Indonesia, oxides (O_x), hydrogen sulfide (H_2S), ammonia (NH_3), and carbon monoxide (CO) are also monitored. It was reported that the majority of air pollution sources of the country are land transportation, industrial emissions, and a densely populated residential (Tri-Tugaswati, 1993; Achmadi, 1996).

4.1.2 Atmospheric Factors

In studying the air pollution problems and their impact on public health, it should be borne in mind that some considerations of the dynamics of the pollutants in the environment were essential (WHO, 2014). Other than the origin of the air pollution, there are some factors not to be taken lightly, namely,

1. The wind velocity and direction are particularly important when developing an urban planning such as the site of industrial parks, settlements, and other sites, since it is expected that the wind will drive away any pollutant from those vicinities. In Indonesia, geographically it was good that most of the major cities where so much pollution was generated are located in the coastal areas of the islands, with the wind blowing throughout the years.
2. Relative humidity of the ambient air is also a factor need to be taken into consideration. Based on the existing data in most cities, Indonesia has an average of relative humidity in the range of 70%–80%. Due to this fact, there are conditions that may worsen the air pollution problems since the presence of so much water vapor in the air may boost the transformation of sulfur oxides (SO_x) to sulfuric acid (H_2SO_4), a chemical corrosive and dangerous to properties and health. The relatively high air temperature in the tropic also intensifies the speed of chemical reactions in the air.
3. Solar radiation in the tropical area plays an important role in transforming hydrocarbons and nitrogen oxides into photochemical oxidants, including ozone (O_3) with its damaging effects on materials (Figure 4.1).

4.2 SOURCES OF AIR POLLUTION IN INDONESIA

From the many sources of air pollutions in Indonesia, a simple and broad category chosen is the natural and anthropogenic sources.

4.2.1 Air Pollution from Volcanoes

One of the most important natural causes of air pollution in Indonesia is from the volcanic eruption. Located in the Pacific Rim, the country lies in the ring of fire in the Asia Pacific region and has the capacity of one-third of the global volcanic activities. The most populous Java Island is also home to 45 active volcanoes. Emissions from volcanic eruption mostly contain SO_2, carbon dioxide, and hydrogen fluoride, together with other volcanic gases and ashes. Volcanic ashes can move hundreds to thousands of miles downwind from a volcano to neighboring cities. Fresh volcanic ashes are gritty, rough, and at times corrosive. The ash may also contain some heavy metals, such as cadmium, mercury, and organic compound polycyclic aromatic hydrocarbons (PAHs) (Symonds et al., 1994; Harada et al., 2013; Abishek and Jeremy, 2009). As a whole, these ashes can cause respiratory problems among young children, the elderly, or those already with respiratory ailments.

The most abundant gas typically released into the atmosphere from volcanic activity is water vapor, followed by carbon dioxide and SO_2. Other smaller amounts

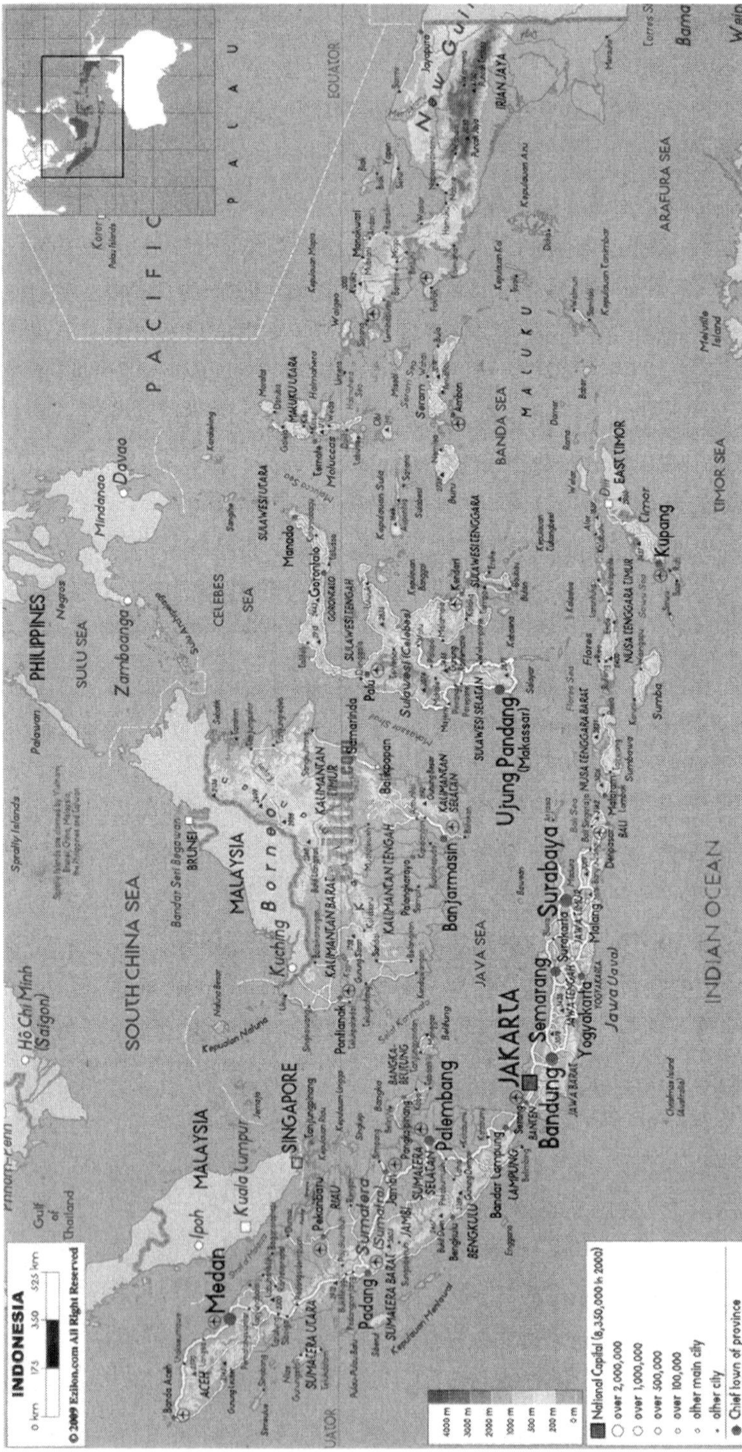

FIGURE 4.1 Map of the Republic of Indonesia with 17,508 islands.

of gases released by volcanic activity include H_2S, hydrogen, CO, hydrogen chloride, hydrogen fluoride, and helium (Symonds et al., 1994).

Materials emitted from a volcanic eruption may also absorb light that reduces visibility and lower the ambient air temperature. The eruption of Mount Merapi in Central Java (2012), Mount Sinabung of Sumatra in (2013), and Mount Kelud in East Java (2014) are examples of these phenomena (Ministry of Environment, 2012).

4.2.2 Air Pollution from Forest Fire

Other natural and partly anthropogenic air pollution is caused by forest fire. Indonesia conserves the tropical rain forests for the benefits of the world. Nonetheless, nearly every year in some places, there have been constant forest-fire episodes, especially in the dry season, for instance, in Sumatra and Kalimantan Islands. The main cause of this forest fire was from natural fire of the self-burning *peat*; but lately the forests were intentionally burnt by some local and international investors in transforming the forest into plantation projects.

The smoke from these forest fires caused considerable air pollution throughout the small and medium cities nearby as well as in neighboring countries. All those forest fires emit CO, SO_2, nitrogen dioxide (NO_2), O_3, and particulate matters. CO, which is a poisonous gas, was emitted in large amounts during forest fires (International Forest Fire News, 2000).

Though the magnitude of suddenness of volcanic eruptions and forest fires as natural causes for air pollution can sometimes be drastic, they are not as damaging as the air pollution caused by human activities. The use of fossil fuels for transportations, industries, power stations, and the like are more serious problems due to their continuous occurrence and insidious health consequences.

It is worth to note here that no particular record of other air pollution with significantly serious impacts such as caused by pollen dispersal, volatile organic compounds (VOCs), or natural radioactivity was reported in Indonesia.

4.2.3 Anthropogenic Source in Urban Area

Anthropogenic sources of air pollution are mainly from motor vehicles, industries, households, even from simple personal activity such as smoking. From the transportation sector, CO is the most well-known air pollutant. This toxic chemical agent is the result of an incomplete combustion of carbon in fossil fuels. CO productions usually increase during traffic jams. It is not only CO; other pollutants are also released into the air such as NO_x, particulate matter of 2.5 microns (μm) and less ($PM_{2.5}$), SO_2, and H_2S. Their presence through emission from the vehicle depends on the amount and quality of fuel used, whether the fuel contains a high or low concentration of sulfur (National Institute of Health Research and Development [NIHRD], 2007; Achmadi, n.d.).

According to surveys conducted by Ministry of Environment (MoE) of Indonesia, there are two large cities that show high concentration of ambient CO and another city of high NO_2 concentration. The results of CO measurements in 22 cities (2011), except in 3 cities, show that the ambient concentration trend is decreasing, an indication of quality improvement of the air. However, NO_2 ambient air concentrations show an increasing trend in 15 cities of the 22 cities evaluated in 2011 and 2012.

Some pollutants in the air will be transformed into other more toxic compounds; for example, SO_2 at certain humidity transforms into sulfite and sulfate compounds that are highly acidic and corrosive (Figures 4.2 and 4.3).

Air pollution in Indonesia is one of the major determinants of the quality of environment, especially in large cities. Transportation contributes about 60%–80% of the air pollution (Tri-Tugaswati, 1993; Achmadi, 1996). Effects of the air pollutants on human health depend on the characteristic of the material.

4.3 PUBLIC HEALTH STUDIES FOR AMBIENT AIR POLLUTION

Outdoor air pollution has always been a major environmental health problem affecting everyone in the world. WHO estimates that some 80% of outdoor air pollution–related premature deaths were due to ischemic heart disease and strokes, while 14% of deaths were due to chronic obstructive pulmonary disease (COPD) or acute lower respiratory infections; and 6% of deaths were due to lung cancer (WHO, 2014).

The problem of air pollution in the island of Java indicates a direct relationship with the rapid increase of population as well as its socioeconomic contribution over the past few decades. It was recorded that in 1960, the population of Jakarta was just 1.2 million. In 2004, it increased to 8.8 million. Later, a more detailed survey estimation showed that in fact the population of greater Jakarta together with the four municipalities surrounding the city has swollen to 23 million in the same year. The recent rapid industrialization and economic growth in Jakarta and some neighboring cities had significantly urbanized a large number of populations into the cities. As a result, overcrowding in the city was faced by those cities. The city of Jakarta itself has a population concentration of more than 32,000 people per square mile (Bureau of Statistics Indonesia, 2014).

The impact of air pollution on human health is very wide, ranging from the local to systemic effects. Lung is the main target organ of air pollutants. Several known lung disorders are asthma, bronchitis, pneumonia (some pollutants affect the body's defense system that weakens local airways), and COPD. Further, some air pollutants (CO, No_x, and O_3) also have their effects on the cardiovascular system. To some extent, air pollutants like Pb, CO, and solvents may also have an impact on the nervous system (neurological effects). Other vital organs that can be affected by air pollutants are the liver and kidney.

Of all the air pollutant, particulate matter is the most influential air pollutant on health in the urban area. It is a complex mixture of very small solids and liquid droplets made up of a number of components, including acids (such as nitrates and sulfates), organic chemicals, metals, and soil or dust particles (United States Environmental Protection Agency [USEPA], 2014). Depending to its size, particulate matter can be inhaled into the respiratory system.

The content of these micro particles is highly source-dependent. Particulates are generated from various sources, including industrial processes, grinding, crushing, volcanic eruptions, volcanic dust, tire frictions, burning carbon fuels, mold spores, and viruses. The sizes of these ultra-small particles determine their potential health problems.

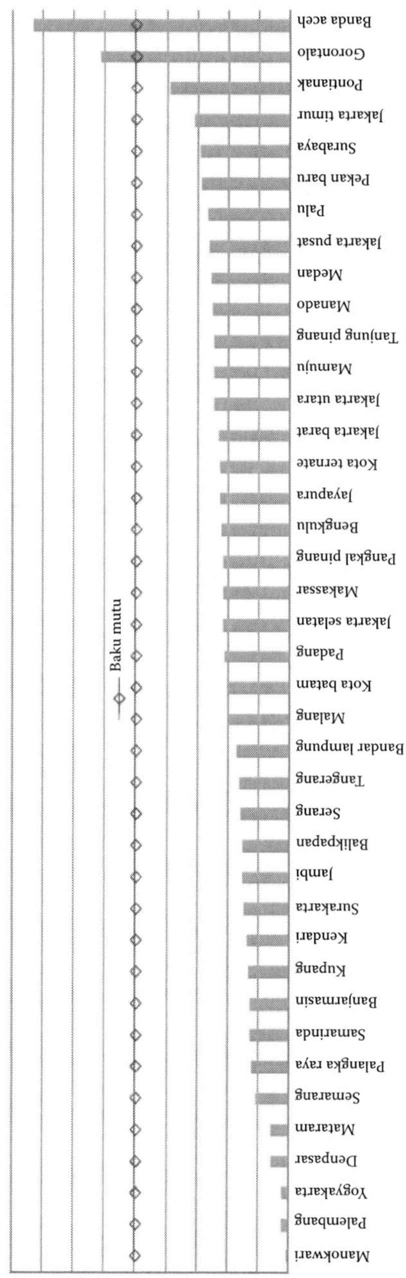

FIGURE 4.2 Concentration of carbon monoxide in big cities in Indonesia 2012 (μg/m^3). (From Ministry of Environment, *Evaluasi Kualitas Udara Perkotaan*, Jakarta, 2012.)

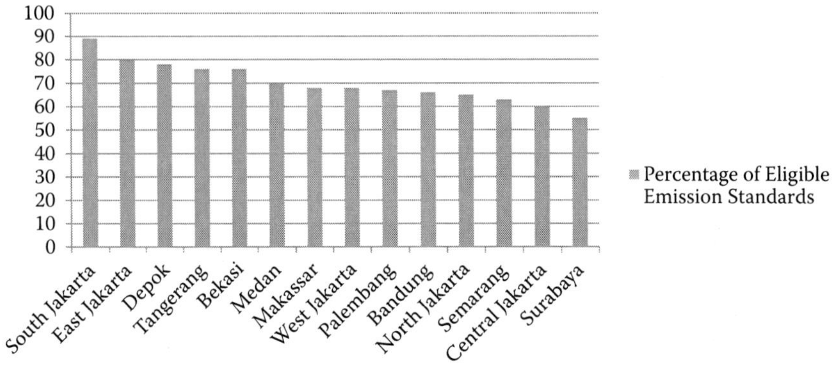

FIGURE 4.3 Percentage of eligible emission standards from transportation in big cities in Indonesia 2012.

For the purpose of navigation and transportation safety, particulate matter was also measured as total suspended particulate (TSP), an indicator of visibility. These micro particles will be suspended in the air; hence they are also called the SPM and become part of the air pollutant other than gases. TSP has a very complex mixture of various organic and inorganic compounds which have size less than 1 μm to 500 μm. In relation to health, USEPA grouped particle pollutants into two categories: first, inhalable coarse particles, such as those found near roadways and dusty industries; their diameters range from larger than 2.5 μm to smaller than 10 μm. The second are fine particles such as those found in smoke and haze, with the size of 2.5 μm in diameter and smaller (USEPA, 2014). For practical reasons, conversion from TSP to particulate matter with the diameter of 10 μm or less (PM_{10}) was sometime done with the equation of PM_{10} concentration equal to 55% of TSP concentration; that is, 100 μg/m^3 of TSP should be equal to 55 μg/m^3 PM_{10} (WHO, 2014).

Particulate matters affect more people than any other pollutant. The major components of particulate matters are sulfates, nitrates, NH_3, sodium chloride, black carbon, mineral dust, and water. It consists of complex mixture of solid and liquid particles, both organic and inorganic substances, suspended in the air (WHO, 2014).

The proven health-damaging particles are those within a group of PM_{10} and easily found in busy cities. While larger particles will mostly stick at the upper respiratory tract, these minute particles can penetrate and lodge deep inside the lungs. Included in this group, $PM_{2.5}$ may even infiltrate deeper into the lung up to the alveoli, making serious health effects such as asthma and COPD. $PM_{2.5}$ contains mostly free silica, radioactive material, and viruses and is also associated with high infant death rate (Cakmak et al., 2009).

The result of the monitoring during a decade ago already indicated that SPM and NO_x in Jakarta were the predominant pollutants (Tri-Tugaswati, 1993; Achmadi, 1996). Other than those, pollutants such as SO_x can react with other compounds in the atmosphere to form small particles. These particles will deeply penetrate into sensitive parts of the lungs and can cause or worsen respiratory disease, such as emphysema and bronchitis (WHO, 2014; USEPA, 2014).

There is a close, quantitative relationship between exposure to high concentrations of small particulates (PM_{10} and $PM_{2.5}$) and increased mortality or morbidity, both daily and over time. Conversely, when concentrations of small and fine particulates are reduced, related mortality will also go down, presuming other factors remain the same (WHO, 2014). So far there are no such extensive correlation studies between air pollution and the health of the community in Indonesia. A study was done by Ostro (1994) on asthma attacks and bronchitis, diseases that are commonly faced by the residents of Jakarta. Air pollutants in Jakarta caused, among other illnesses, approximately 1200 cases of premature mortality, 32 million cases of respiratory symptoms, and 464,000 cases of asthma attacks. Respiratory hospital admissions (RHAs) due to ambient sulfate and TSP levels have also been continually rising in Jakarta due to severe air pollution.

Based on epidemiological and experimental laboratory study, it was verified that elevated concentrations of particulate air pollution contribute to cardiovascular morbidity, hospitalization, and mortality (Verrier et al., 2002). WHO (2014) also stated that reduction of annual average PM_{10} concentrations from levels of 70 μg/m^3 (common in many developing cities) to the WHO guideline level of 20 μg/m^3 could reduce air pollution–related deaths by around 15%.

The Center for Health and Ecology Research and Development, NIHRD, Ministry of Health (NIHRD, 2007) has 25 air pollution sampling points in five sites of Jakarta including in West Jakarta, North Jakarta, Central Jakarta, East Jakarta, and South Jakarta. Some parts of the city showed TSP level of 350 μg/m^3, more than 70 μg/m^3 the permissible level of TSP set in the WHO guidelines. All sampling points showed TSP concentration ranged between 74.07 μg/m^3 and 416.26 μg/m^3. The highest TSP concentrations were found in busy places of Jakarta, that is, Central Jakarta, where offices are located, and East Jakarta, the industrial area. Meanwhile lead concentrations at some sampling points showed concentration levels in the range of 0.00 μg/m^3–3.88 μg/m^3.

Particulate matter was always a problem in many cities in Asia. For example in Nanjing, China, the concentration level of TSP on both sides of a highway was six times of the national limit value (Zhao and Shi, 2012).

Every air pollution substance has its own characteristic in causing the ill effect on human health. Serious risks to health were seen from NO_2, O_3, and SO_2. NO_2 and SO_2 can play a role in asthma, bronchial symptoms, lung inflammation and reduced lung function. Epidemiological studies have shown that symptoms of bronchitis in asthmatic children increase in association with exposure to certain concentration levels of NO_2. Reduced lung function growth is also linked to NO_2 at concentrations found in cities of Indonesia, particularly in the island of Java (NIHRD, 2007).

Excessive O_3 in the air can have a marked effect on human health. O_3 is known as a major factor in asthma morbidity and mortality. It can cause breathing difficulty problems, reduce lung function, and cause lung diseases. In Europe, it is currently one of the air pollutants of most concern (Grundstrom and Pleijel, 2014). At the ground level, O_3 is one of the major constituents of photochemical smog. By the help of sunlight acting as catalytic substance (photochemical reaction), O_3 is formed through the reaction of molecular oxygen with atomic oxygen released by NO_2. In the daytime, primarily at busy hours of the city, much of NO_x and VOCs are emitted by vehicles, solvents, and industry. Therefore, high levels of O_3 pollution usually occur during periods of sunny weather. NIHRD (2007) found that Jakarta had high concentrations of nitrogen oxide, ranging between 23.61 μg/m^3 and 55.36 μg/m^3.

SO_2 is a colorless gas with a sharp odor. The main anthropogenic source of SO_2 is the burning of sulfur-containing fossil fuels (coal and oil) for domestic heating, power generation, motor vehicles, and the smelting of mineral ores that contain sulfur (WHO, 2014; USEPA, 2014). In Indonesia, its main source is from fossil fuels in the urban area as well as natural sources like forest fires and volcanic eruptions occasionally (Ministry of Environment, 2012).

As a pollutant SO_2 can affect the respiratory system, functions of the lungs, causes irritation of the eyes, and makes people prone to infections of the respiratory tract. Inflammation of the respiratory tract causes coughing, excessive mucus secretion, aggravation of asthma, and chronic bronchitis. Hospital admissions for cardiac disease and mortality increase on days with higher SO_2 levels. Irritation of the nose and throat will be seen immediately at an exposure to a concentration of 6–12 ppm SO_2. For limits of exposure, The WHO recommends a concentration of no greater than 0.008 ppm (20 μg/m^3) 24 hours mean (WHO, 2006). When SO_2 combines with water, it forms H_2SO_4: the main component of acid rain caused by forest fire and/or the burning of fossil fuels.

Studies on a direct link between SO_2 and asthma incidence in cities of Indonesia are scarce, but surveys of ambient air pollutant in some cities done by NIHRD (2007) indicated that concentrations of SO_2 have exceeded the permissible standard. Some evidences showed an association between short-term exposure to SO_2 and increased visits to emergency departments and hospital admissions for respiratory illnesses, particularly among the at-risk populations like children, the elderly, and asthmatics (USEPA, 2014). Studies indicate that a proportion of people with asthma experience changes in pulmonary function and respiratory symptoms after periods of exposure to SO_2 as short as 10 minutes.

Another finding important to health is the ability of SO_x to react with other compounds in the atmosphere to form small particles. These particles will then penetrate deeply into sensitive parts of the lungs and potentially cause or worsen respiratory disease, such as emphysema and bronchitis, and can aggravate existing heart disease, leading to increased hospital admissions and premature death (USEPA, 2014).

4.3.1 Public Health Studies for Natural Sources of Air Pollution in Indonesia

4.3.1.1 Volcano

One of the most active volcanoes among 45 of volcanoes in the island of Java is Mount Merapi. This volcano is the youngest and most active volcano, with its height of 2968 meters above the sea. Mount Merapi is surrounded by cities, that is, Jogjakarta with its 2 million population, located 30 kilometers from the mountain and Magelang, which has approximately 1 million population, and the small town of Boyolali, which has around 500,000 population, located in the eastern side of the volcano. Mount Merapi is located in relatively the middle of the island of Java and occupied by villagers on its foot as close as 4 kilometers from the cap.

During the eruption, there were 434,699 people displaced. Some clinical adverse effects were observed among people visiting the emergency clinic at the nearby camp. These people mainly suffered from upper respiratory infections (Figure 4.4). In their situation, anxiety leading to psychological stress was also seen among the displaced people.

The latest eruption of Mount Merapi occur in September 2010; since then several eruptions took place nearly every day and lasted for 3 months. The ash reached the neighboring towns and cities in Java up to 200–400 kilometers away (e.g., up to the far reach of Bandung and Bogor; two cities lay close to Jakarta).

Four points of monitoring stations were set for the air pollutants monitors located 20–25 kilometers from the active Mount Merapi. These monitoring stations were located at the four sides of the volcano, 11 kilometers south of the center of Jogjakarta City. At these monitoring stations, ashes were usually visible to the naked eyes.

1. Eastern side from the volcano: at the Jurang-Rejo Klaten Village, ashes were not easily visible to the naked eyes, relative humidity 70%.
2. Western side from the volcano: at the center of Magelang City ashes were visible to the naked eyes. In addition, monitoring was also done close to the Borobudur Temple, the popular world's heritage, relative humidity 50%.
3. Northern side from the volcano: at Sidomukti area in Salatiga City, relative humidity 90%.
4. Southern side from the volcano: in Klaten, a small city, relative humidity 70%.

The western part of Mount Merapi, that is, the city of Magelang, suffered the worst concentration of ashes, with a concentration as high as 1609.10 μg/m^3. Table 4.1 shows the temporary monitoring results of air quality around Mount Merapi (20–25 kilometers northern, southern, eastern, and western sides of Merapi).

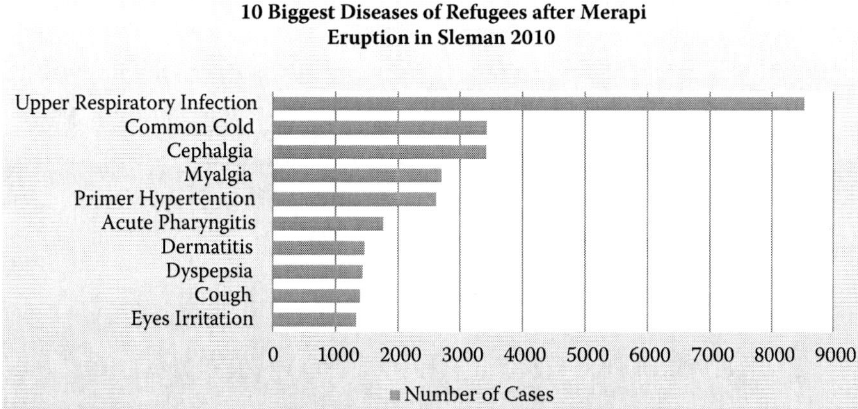

FIGURE 4.4 Distribution of complaints of the displaced people visiting the emergency camp clinic near the volcano. (From www.merapicombine.or.id.)

TABLE 4.1
Analyses of Pollutant Mount Merapi Eruption 2010

	Location							
Parameters	Salatiga City	Klaten City	North of Jogjakarta City	Regional Office Ministry of Environment	Center of Magelang City	Gulon Village	Borobudur Temple	Bowan, Delanggu Village
TSP (μg/M³)	63.1	70.8	136.7	163.2	1609.10	2019	567.4	43.2
SO_2 (μg/M³)	11.4	8.7	26.1	16.6	5.6	5.7	9.2	Not analyzed
H_2S (ppm)	0.0003	0.0001	0.0001	0.0001	0.0016	0.0009	0.0003	Not analyzed
NO_2 (μg/M³)	3.7	3.9	8.2	6.7	10.6	6.3	6.8	Not analyzed

Source: Ministry of Environment, *Evaluasi Kualitas Udara Perkotaan,* Jakarta, 2012.

Referring to the TSP levels of different places in Table 4.1, the ambient air quality monitoring analysis above shows that the dominant bursts of volcanic ash moved to the south and west, with concentrations of TSP as high as 1609.10 μg/m³ at the center of Magelang City and 2019 μg/m³ at Gulon Village. These places show concentrations of TSP that seriously exceed the quality standards of the National Regulation No. 41 of 1999 on ambient air quality.

4.3.1.2 Forest Fire

Forest fire–related air pollution episodes are a recurrent phenomenon in Southeast Asia. Nine such incidents have been reported over the last 20 years, of which the 1997/1998 smoke haze episode attracted the broadest attention. The air pollutant that predominantly had health impact was particulate matter.

Eye irritation and the increase of incidence of upper respiratory infection were reported in the city of Pekanbaru, Sumatera, during the forest fire episodes (Ministry of Health, 2013). Besides causing acute and chronic respiratory diseases such as bronchitis, acute asthma, and upper respiratory tract infections, due to its capability of scattering and absorbing light, particulates also reduced visibility with the result of disturbing transportation.

From forest fires, mixtures of soot, tars, and volatile organic substances, in the form of particulates either solid or liquid, are emitted in large quantities. As is well known, particulates smaller than 2.5 μm in diameter have the potential of being deeply inhaled into the lungs and penetrating the alveoli. It damages the lung tissues, causing respiratory and cardiovascular problems. Meanwhile, SO_2 emissions are negligable in forest fires. Most forest fires contain SO_2 concentrations of less than 0.2%. If the temperature of a forest fire was higher than 1500°C, like in a very severe forest fire, nitrogen oxides may be released in significant quantities.

4.4 PUBLIC HEALTH STUDIES FOR INDOOR AIR POLLUTION

With the ambient air being polluted in major cities, indoor air pollution is also a health risk problem seen in the country. The problem is commonly found in urban areas as well as in the rural areas.

Major cities with the increasing number of high-rise buildings supported with air conditioning have their potential tolls on health. Rooms in these buildings pose health threats from physical, biological, and chemical contaminants. They are usually equipped with furniture made of processed woods or fabrics containing benzene and other organic compounds toxic to humans. Likewise, air conditioning has been known to account for Legionnaire's disease that in some cases may be fatal for the victims. Though no clear data was found, facts of ambient air temperature of the tropics in Indonesia, the design of high-rise buildings, and for the reason of comfort of building occupants, the designers were imposed to use air conditioning to combat the high temperature and air pollution of the outside ambient air. A neglected maintenance of the air-conditioning systems potentially promotes the growth and spread of microorganisms.

Some slum areas in Jakarta had also shown the risks of indoor air pollution. A study run in Pekojan (Purwana, 1999) reported that the high prevalence of respiratory problems among children under 5 years old was statistically related to the PM_{10} concentration of 70 $\mu g/m^3$ and above. Among the sources of the indoor pollutants of the place was cigarette smoke.

Together with its being a problem in urban areas, indoor air pollution is also found in rural areas. Referring to the WHO citation, indoor smoke is a serious health risk for some billions of people who cook and heat their homes with biomass fuels and coal (WHO, 2014).

A research in remote area of Hoineno Village in the Island of Timor (Purwana et al., 2012) confirmed a statistical relationship between acute respiratory infection (ARI) and PM_{10} concentration in traditional huts of the natives (Figure 4.5). The major source is smoke from wood-burning stoves.

For their daily cooking and warming, the natives have a simple three-stone stove burning firewood nearly all day long that produced abundant smoke in the huts. No effective ventilation was installed in the huts. The only access to the inner part of the hut is through a small opening, the size of less than a meter square (0.7×1.0 m^2), as an entrance. The hut itself is made of dried straws of tallgrass. The single room of huts accumulates the smoke indicated by the high level of CO up to a maximum of 240 ppm (Purwana et al., 2013).

Known as a toxic gas, together with other toxic materials CO is produced from an incomplete oxidation of carbons given off by burning of wood. Its stronger affinity to hemoglobin in blood reduces the capacity of the blood to carry oxygen. The effect of all these is a decrease of oxygen supply into the cells of the body and most important is the impact on brain cells that have the highest need of oxygen supply.

To solve the problem of indoor smoke in this community was not simple. Smoke is also used for the preservation of corn, the staple food. Keeping smoke inside the hut is important for that purpose. No alteration of the hut (like applying windows for ventilation) is permitted by tradition. At present step, a team from the Universitas Indonesia applied a partition and a chimney inside the traditional huts that help to drive the smoke to the open air. It is still too early to claim that this action will directly reduce the prevalence of ARI since other health risks are also endemic in the place; among others are malaria, undernutrition, diarrheal diseases, and tuberculosis.

FIGURE 4.5 A typical traditional hut in Hoineno Village, Timor, Indonesia.

Further action research on this problem is on the way (Purwana et al., 2012). It is worth noting that in the research place the health problems intermingle with the social and economic problems.

4.4.1 Risk Assessment of Cement Industry Contribution: A Case Study

Indonesia is a country with a large cement industry. In 2014, cement production was 68 million tons, an increase from the 60 million tons production in 2013. Total production is expected to increase to 90–95 million tons in 2017 (Ministry of Industry, 2013). This figure is quite large so it requires considerable energy support.

The cement industry in Indonesia spread in several cities and regions with growing production seen every year. With the increase in production, it increases the need of more fuel. In 2013, the Ministry of Energy and Mineral Resources (Ministry of Energy and Mineralogy, 2013) estimated that Indonesian domestic will need 95.55 million tons of coal in 2014, with the largest allocation of 57.4 million tons to the national power company and 9.8 million tons to the cement industry.

Coal is not the only source of air pollution. The cement industry in a way has its distinct role as one of the largest sources of air pollutants. The cement industry emits many kinds of pollutants into the air such as SO_2, CO, silica, and particulates. Wiguna (2006) stated that particulate matter from industrial furnace is the largest contributor to air pollution (Okta, 2006). Industrial furnaces, including from the cement industry, contribute 51.27% of total particulates air pollutant emissions and accounts for 5% of global CO_2 emissions (Zeleke et al., 2010). In some areas, cement industries are the major source of air pollutants. Studies on the level of air pollution in the industrial sector of South Sumatra in 2011 showed that the cement industry is one of the largest producers of the air pollution load. Other than its impact on environmental degradation, air pollution has also its indirect impact on the economic sector. Indirectly, environmental pollution will play as an obstacle to the development of Indonesia's economy after the struggle to get out of the economic crisis in the past and the country's effort to build a healthy sustainable economy.

The development of cement industry in the context of health impact was studied. Based on studies of air pollution emitted by industries, the cement industry was prioritized to prevent air pollution. The priority level was calculated based on the amount of emissions released by the plant and cleanup costs to be incurred to address the contamination (Budi, 2000).

Particulates emission from cement industry contains hazardous substances such as tricalcium silica, dicalcium silica, alumina, iron oxide, CO, SO_2, and other substances (Mwaiselage et al., 2005). These materials are toxic materials and may cause irritation of the gastric mucosa, the mucosa of the lungs, skin disorders, respiratory disorders, and cancer (Al-Neaimi et al., 2001).

These health risks are also may apply for the people around the plant to bear. For some reasons (among others, weak regulation enforcements), in many regions in the country, a lot of people flock closely near the cement industry plants. Many studies have been done to estimate the public health impact of particulate matter exposure emitted by cement industry. A research on risk assessment of exposure to ambient air particulate to communities in Padang, West Sumatra around cement industry proposes that people living in the radius of 2.5 kilometers and less from the center of the industry have a substantial risk to respiratory disorders (Novirsa and Achmadi, 2012).

The figure showed that it exceeded the national ambient air quality standards (NAAQS) (35 $\mu g/m^3$) established by the USEPA. According to the standards, there were three areas that exceed the 24-hour exposure standard, namely the site at 1 kilometer away from the source, that is, 41 $\mu g/m^3$, at the distance of 2.5 kilometers (38 $\mu g/m^3$), and of 3 kilometers (37 $\mu g/m^3$) (Al-Neaimi et al., 2001).

It was found that the risk area of particulates exposure was detected at radiuses of 500–1000 meters, 1500–2000 meters, and 2000–2500 meters. The risk is directly proportional to the concentration levels of particulate and exposure levels. The greater the exposure levels the greater the risk of the area. Similar study conducted by Daud (2010) revealed that the risk of SO_2 in an industrial area of South Sulawesi was higher when the distance got closer to the industrial area (OR = 5.83) (Anwar, 2010) (Figure 4.6).

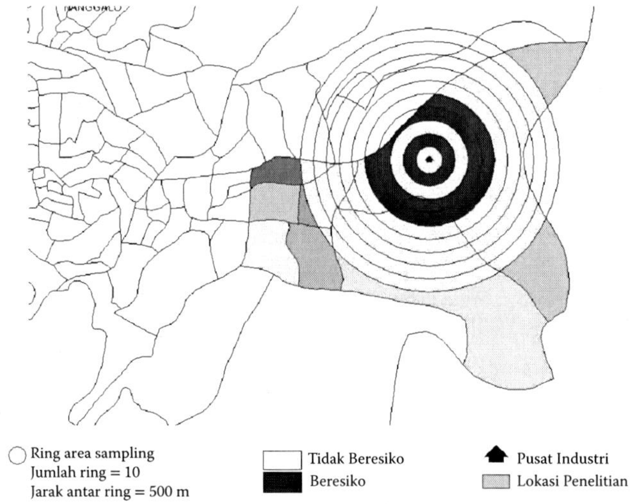

FIGURE 4.6 Risk area of particulate exposure to residents around cement industry plant, West Sumatera 2012.

4.4.2 Thermal Inversion: The Case of Bandung City

Open air pollution is closely linked to climate and topography. Air pollution episodes can be particularly troublesome if an affected city is located in a valley surrounded by high grounds. This was the case in the Meuse Valley in Belgium and is the case in Mexico City, Mexico (WHO, 2014). Thermal inversion can occur in a place if the place or a city is located in a relatively deep valley surrounded by hills. In the morning, when there is no considerable wind draft, human activities begin issuing pollution materials such as from automobile, industrial, and household activities, while the sun just starts to rise from behind the hills and generates a hot layer of air above the valley. In this case, the normal vertical temperature gradient is inverted in that the air is cooler near the surface of the earth. This less-dense hot air mass resting on top of the cooler and denser layer of air mass will obstruct a variety of air pollutants trapped in the valley due to no upward movement of the cooler air layer from the valley. Thus a concentration of pollutants remains settled at the ground level in the valley with its all human daily activities. As a result, the population will be trapped in relatively high air pollutants due to low wind movement. An inversion can also lead to smog being trapped close to the ground, with possible adverse effects on health. This type of inversion occurs in the vicinity of warm fronts. With sufficient humidity in the cooler layer, fog is typically present below the inversion cap. In cities where many pollutants exist, this can be a serious problem.

Due to the topographical position of an area, air inversion phenomenon usually occurs in some big cities such as Los Angeles City and Mexico City in the United States, Mumbai in India, and also Bandung in Indonesia. The greater part of the city of Bandung lies on the flat area of 2000 meters above the sea level. It was a huge caldera in the ancient years. At present, it forms a relatively plane topographical ground surrounded by chains of hills and mountains. Due to this topographical shape together with inversion of the air within, the city acts like a bottle with polluted air inside. So far there is no study associated to this phenomenon.

4.5 ECONOMIC BURDEN AND INTERVENTION

Since 1996, the MoE has been implementing the Blue Sky/Clean Air Program (Prodasih) to control the air pollution problem from mobile and stationary sources in particular cities. The Blue Sky Program at the central government level is coordinated by MoE and is implemented in every regency/city with the assistance from the governor/provincial government. Four provinces (The greater Jakarta, West Java, Central Java, and East Java) were designated as a priority for the Blue Sky Program in 1996. It also envisages aggressive social campaigns to bring environmental awareness at a national level.

Jakarta currently is improving considerably its public transportation system to check air pollution. Buses are the chief mode of public transport. A public transport system could relieve Jakarta from air pollution. The idea of a monorail now is in progress. A "Car-Free-Day," which means no cars allowed entering a particular area as well as policy of "Three-in-One" (meaning only cars with the number of passengers of three or more are allowed to enter certain boulevards during the busy hours) (Gunawan, 2008). However, despite the Blue Sky Program, Jakarta sees severe traffic jams almost every day as the number of private cars on the capital's streets continues to rise, with the city seemingly running out of options to deal with the problem. The cost estimates associated with air pollution can be calculated based on the loss to health and nonhealth. An extensive study of health cost was done in 1994 (Ostro, 1994). Ideally valuation of health effects should include the costs of illness such as medical costs, income lost, averting expenditures, and the less tangible effects of illness on well-being such as short of breath, cough, discomfort, and restrictions during the no-work-activities period. Health impacts valued by willingness to pay (WTP) incorporate all of these impacts, whereas a cost of illness (COI) approach only includes expenses such as medical costs and income lost. This way of valuation excludes the economic costs of mortality. An example of this valuation is RHA, assuming that the average stay in the hospital of 10 days plus for the lost working days based on the wage rate. The respiratory health admission can be calculated for each person. The total can be calculated to the number of people admitted into hospital, assuming that the admission is directly linked to air pollution. Firdaus Ali (2014), an environmental expert from the University of Indonesia, calculated the loss caused by air pollution in Jakarta based on the estimated amount of fuel wasted, lost productive time, losses incurred by public transportation owners, and the health costs resulting from congestion. At the time, traffic congestion in the city made a burden on Jakarta residents as much as USD 28.1 billion annually. The biggest loss was attributed to fuel inefficiency, the fuel wasted when vehicles were trapped in traffic jams, which costs USD 10.7

billion per year. In the second place is the loss of productive time, estimated at USD 9.7 billion per year, followed by health costs of 5.8 billion. Public transportation owners suffer a staggering 1.9 billion loss each year due to lost opportunities as traffic jams drastically limit the movement of public transportation vehicles.

4.6 CONCLUSION

Considering the location in the tropics and being and archipelago country, Indonesia shows a unique air pollution problem among Asian countries. The country consists of relatively small islands, with their different variables that may influence the nature of air pollution problems. Meanwhile, 60% of the total population live in Java Island, a relatively small island which accounts for only 7% of the total land mass. From this fact, it is obvious that air pollution in Java Island is strongly related with the social and economic development. Therefore Java Island may be considered burdened by the anthropogenic air pollution.

In addition to anthropogenic sources, Indonesia has natural sources of pollutant from volcanoes and forest fire. Though Java Island is the smallest island among the five major islands in Indonesia, it has one-third of the total number of volcanoes of the country. Some of the active volcanoes may erupt unpredictably. Although the eruptions are not very frequent, during episodes of eruption, the pollutant had quite a significant impact on the health of the people as well as a potential impact to global heritages, for example, Borobudur Temple. Such problems may also add the existing problem of air pollution in urban areas which is mainly located in Java Island.

Other mid-sized cities, such as cities in Sumatra, Kalimantan, and other islands, are also having air pollution problems. Air pollution in these places is primarily due to forest fire and other natural sources with not too distinct an urban air pollution problem, compared with the air pollution problem in Java Island. An example of indoor air pollution has also been described, showing a problem of air pollution linked to traditional culture in Timor Island. Although the country has the program to eliminate air pollution, it seems that the country is still creeping in dealing with air pollution.

REFERENCES

Abishek, T. and Jeremy, C. (2009). *Air Pollution: Measurement, Modelling and Mitigation*, 3rd edition. Taylor and Francis e-Library. Available: http://www.amazon.co.uk/Air-Pollution-Measurement-Modelling-Mitigation/dp/0415479320#reader_0415479320

Achmadi, U. F. (n.d.). *Dasar Dasar Penyakit Berbasis Lingkungan* (*Eco-perspectives of the Disease Occurrences*), 5th edition, Jakarta: Rajagrafindo.

Achmadi, U. F. (1996). Public health implications of environmental pollution in urban Indonesia. *Asia Pac J Clin Nutr*, 5, 141–4.

Achmadi, U. F. (2010). Public health perspectives of small island communities in Indonesia: Issues and challenge. Asia Pacific Public Health Conference. Denpasar, Bali Indonesia.

Al-Neaimi, Y. I., Gomes, J., and Lloyd, O. L. (2001). Respiratory illnesses and ventilatory function among workers at a cement factory in a rapidly developing country. *Occup Med (Lond)*, 51, 367–73.

Ali, F. (2014). Jakarta traffic costs public $2.8b per year, says expert [Online]. Available: http://www.thejakartaglobe.com/archive/jakarta-traffic-costs-public-28b-per-year-says-expert.

Anwar, D. (2010). Analisis cluster terhadap tingkat pencemaran udara pada pabrik cement di Sumatera selatan. *Jurnal Kesehatan Masyarakat Universitas Hasanudin*, 5.

Budi, R. M. (2000). Air pollution and water pollution in the industry of Indonesia. *Environmental Economic*, 47, 73.

Bureau of Statistics Indonesia. (2014). *Statistic of Indonesia*. Jakarta.

Cakmak, S., Dales, R. E., and Vida, C. B. (2009). Components of particulate air pollution and mortality in Chile. *Int J Occup Environ Health*, 15, 152–8.

Grundstrom, M. and Pleijel, H. (2014). Limited effect of urban tree vegetation on NO_2 and O_3 concentrations near a traffic route. *Environ Pollut*, 189, 73–6.

Gunawan, N. K. (2008). Polusi udara akibat aktivitas kendaraan bermotor di jalan perkotaan pulau jawa dan bali. *Pusat Litbang Jalan dan Jembatan*, 1–13. Available: http://pu.go.id/uploads/services/infopublik20130926120104.pdf

Harada, I., Yoshii, Y., Kaba, Y., Saito, H., Goto, Y., Alimuddin, I., Kuriyama, K., Machida, I., and Kuze, H. (2013). Measurement of volcanic SO_2 concentration in Miyakejima using differential optical absorption spectroscopy (DOAS). *OJAP*, 2, 36–46.

International Forest Fire News. (2000). The 1997–98 air pollution episode in Southeast Asia generated by vegetation fires in Indonesia [Online]. Available: http://www.fire.unireiburg.de/iffn/country/id/id_32.htm 23.

Ministry of Energy and Mineralogy. (2013). Kebutuhan batubara domestik 95.550.000 Ton [Online]. Available: http://www.esdm.go.id/berita/batubara/44-batubara/6396-tahun-2014-kebutuhan-batubara-domestik-95550000-ton.html.

Ministry of Environment. (2012). *Evaluasi Kualitas Udara Perkotaan*. Jakarta.

Ministry of Health. (2013). *Profil Kasus Kunjungan RS Kota Pakanbaru*. Sumatra.

Ministry of Industry. (2013). Kementerian perindustrian republik Indonesia [Online]. Available: http://www.kemenperin.go.id/artikel/1903/Indonesia-Produsen-Utama-Biodiesel.

Mwaiselage, J., Bratveit, M., Moen, B. E., and Mashalla, Y. (2005). Respiratory symptoms and chronic obstructive pulmonary disease among cement factory workers. *Scand J Work Environ Health*, 31, 316–23.

National Institute of Health Research and Development. (2007). *Addressing Ambient Air Pollution in Jakarta, Indonesia*. Indonesia.

Novirsa, R. and Achmadi, U. F. (2012). Risk analyses of PM2.5 exposure in industrial park community West Sumatra. *J of Public Health,* 7, 173–9.

Okta, W. (2006). Polutan industri jangan diabaikan. In: Muhammad, D. and Nurbianto, A. D. (eds.) *Jakarta Kota Polusi Mengguµgat Hak Atas Udara Bersih*. Jakarta: Rajagrafindo.

Ostro, B. (1994). *Estimating Health Effects of Air Pollutants: A Methodology with an Application to Jakarta, Policy Research Working Paper 1301*. Washington, DC: World Bank.

Purwana, R. (1999). Partikulat rumah sebagai faktor risiko gangguan pernapasan anak balita (House particulate as a factor for respiratory symptoms among children under-five; a dissertation). Universitas Indonesia.

Purwana, R., Hermawati, E., and Tahun, O. D. R. (2012). Laporan penelitian: Kualitas Udara Dalam Rumah Masyarakat Adat sebagai Faktor Risiko Kejadian Penyakit ISPA pada Anak Berumur sampai dengan Lima Tahun di Nusa Tenggara Timur, 2012 (Research report: The quality of indoor air of traditional huts as a risk factor to ARI among children under-five in Nusa Tenggara Timur, 2012). Universitas Indonesia.

Purwana, R., Wulandari, R. A., Hartono, B., Tahun, O. D. R., and Abidin, N. (2013). Laporan Penelitian: Penerapan Partisi dan Tungku Bercerobong pada Rumah Bulat di Nusa Tenggara Timur (Research report: The application of partitions and chimney on stoves, in the traditional huts of Nusa Tenggara Timur). Universitas Indonesia.

Symonds, R. B., Rose, W. I., Bluth, G., and Gerlach, T. M. (1994). Volcanic gas studies: Methods, results, and applications. In Carroll, M. R. and Holloway, J. R. (eds.) Volatiles in Magmas: Mineralogical Society of America Reviews in Mineralogy.

Tri-Tugaswati, A. (1993). Review of air pollution and its health impact in Indonesia. *Environ Res*, 63, 95–100.
United States Environmental Protection Agency. (2014). Available: http://www.epa.gov/air/sulfurdioxide/health.html.
Verrier, R. L., Mittleman, M. A., and Stone, P. H. (2002). Air pollution: An insidious and pervasive component of cardiac risk. *Circulation*, 106, 890–2.
World Health Organization. (2006). *Air Quality Guidelines for Particulate Matter, Ozone, Nitrogen Dioxide and Sulfur Dioxide Global Update 2005 Summary Of Risk Assessment*. Geneva.
World Health Organization. (2014). Air quality guidelines [Online]. Geneva. Available: http://www.who.int/mediacentre/factsheets/fs313.
Zeleke, Z. K., Moen, B. E., and Bratveit, M. (2010). Cement dust exposure and acute lung function: A cross shift study. *BMC Pulm Med*, 10, 19.
Zhao, Y. and Shi, D. (2012). Analysis of total suspended particulates pollution along Shanghai-Nanjing expressway. *OJAP*, 1, 31–6.

5 Air Pollution and Health in Malaysia

Mazrura Sahani, Md Firoz Khan, Wan Rozita Wan Mahiyuddin, Mohd Talib Latif, Chris Fook Sheng Ng, Mohd Famey Yussoff, Amir Afiq Abdullah, Er Ah Choy, Norhayati Mohd Tahir

CONTENTS

5.1 BACKGROUND

This chapter contains an overview of air pollution and health in Malaysia. It aims to provide general information about Malaysia and the major cities, past trends and episodes of air pollution, measurement of exposure, and epidemiological studies of air pollution in Malaysia and the air quality management in the country.

5.1.1 General Information about the Country and Cities

Malaysia is located on in the center of Southeast Asia surrounded by the South China Sea, Malacca Straits, and the Sulu Sea. The country is crescent-shaped, starting with

Peninsular Malaysia (West Malaysia) and extending to Sabah and Sarawak (East Malaysia), located on the island of Borneo.

Malaysia covers an area of about 330,803 square kilometers, consisting of 11 states in Peninsular Malaysia—namely Johor, Kedah, Kelantan, Malacca, Negeri Sembilan, Pahang, Perak, Perlis, Penang, Selangor, and Terengganu—and the federal territories of Kuala Lumpur and Putrajaya, Sabah and Sarawak on the island of Borneo, and the federal territory of Labuan off Sabah (Figure 5.1) (Economic Planning Unit [EPU], 2013).

Malaysia lies entirely in the equatorial zone with the characteristic features of the climate of Malaysia, which are uniform temperature, high humidity, and copious rainfall (Malaysian Meteorological Department [MMD], 2015b). The average rainfall is around 250 centimeters (98 inches) a year and the average daily temperature throughout Malaysia varies from 21°C to 32°C. The climates of the peninsula and East Malaysia differ, as the climate on the peninsula is directly affected by wind from the mainland, as opposed to the more maritime weather of the east (MMD, 2015a). Due to the location and its influence by the Pacific Ocean, Malaysia is exposed to the El Niño effect, which reduces rainfall in the dry season. Change of climate is likely to have a significant effect on Malaysia, increasing sea levels and rainfall, increasing flooding risks and leading to large droughts.

Malaysia's 2010 mid-year population is estimated to be 30.34 million. Population structure in Malaysia in terms of the ethnic group is comprised of Malay (50.1%), Chinese (22.6%), Indigenous (11.8%), Indian (6.7%), Others (0.7%), and noncitizens (8.2%). The distribution of the number of urban centers by population size class for the states in Malaysia showed that Selangor by far had the highest number of urban centers in the country, followed by Johor and Perak. In the metropolitan category—that is, population size class of 75,000 persons and above—Selangor

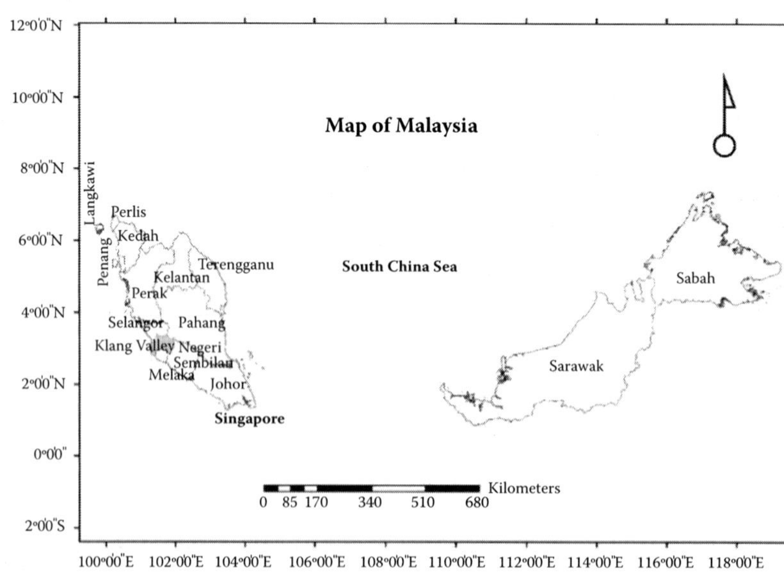

FIGURE 5.1 Map of Malaysia.

topped the list with 10 urban centers. In 2000, all states, with the exception of Perlis and Melaka, had more than half of their urban population in the size class of 75,000 persons and above.

Beginning in 1971 through the late 1990s, Malaysia transformed itself from a producer of raw materials into an emerging multisector economy. Growth was almost exclusively driven by exports—particularly of electronics. As a result, Malaysia was hard-hit by the global economic downturn and the slump in the information technology (IT) sector in 2001 and 2002. Gross domestic product (GDP) in 2001 grew only 0.5% due to an estimated 11% contraction in exports, but a substantial fiscal stimulus package equal to USD 1.9 billion mitigated the worst of the recession and the economy rebounded in 2002 with a 4.1% increase. The economy grew 4.9% in 2003, notwithstanding a difficult first half, when external pressures from severe acute respiratory syndrome (SARS) and the Iraq War led to caution in the business community. Growth topped 7% in 2004. Healthy foreign exchange reserves, low inflation, and a small external debt are all strengths that would minimize the risks of Malaysia experiencing a future financial crisis similar to the one in 1997. The economy remains dependent on continued growth in the United States, China, and Japan, top export destinations and key sources of foreign investment.

Klang Valley is the most urbanized region in Malaysia. This Klang Valley region (Figure 5.2) consists of the federal territory of Kuala Lumpur and the Selangor districts of Gombak, Petaling, Hulu Langat, and Klang and several local authorities, including three important councils: namely Kuala Lumpur City Hall, Shah Alam City Council, and Petaling Jaya City Council. This complex conurbation known as Kuala Lumpur conurbation (KLC) developed historically through the progressive development of satellite towns from Kuala Lumpur, especially after World War II. In the early 1970s, Klang Valley was acknowledged as a coherent urban planning region (Katiman, 1997). Despite its small area of about 242.3 square kilometers, which is approximately 1.25% of the size of Malaysia, the region's population represents

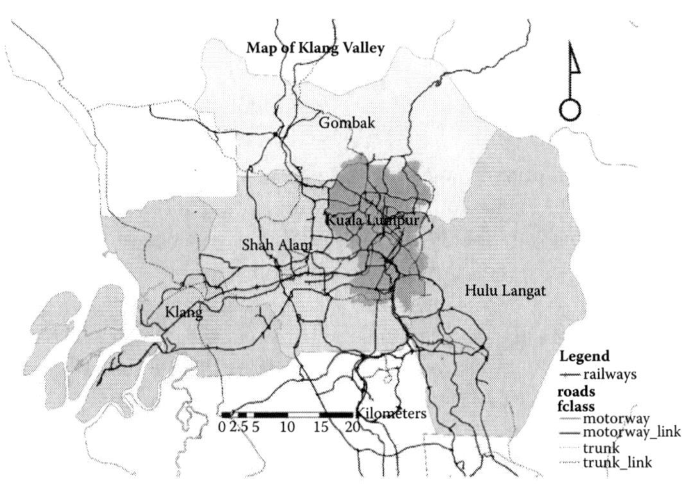

FIGURE 5.2 Map of Klang Valley region with Kuala Lumpur and its conurbation.

17.4% of the national population. Kuala Lumpur, located in the Klang Valley region, comprises the highest density of population, with 5,639 persons per square kilometer. Since the Kuala Lumpur Structure Plan (KLSP) 1984 (KL, 2020), the other urban centers in the Klang Valley region, notably Petaling Jaya, Shah Alam, and Subang Jaya, have grown at a rate that far outstrips that of the city. There has been strong in-migration to the KLC outside Kuala Lumpur from all over the country and net out-migration from Kuala Lumpur into residential areas located outside the city. In the year 2000, the population of Kuala Lumpur was approximately 1.42 million, compared with 4.30 million for the whole of the KLC, a population distribution pattern not envisaged by the KLSP 1984 (KL, 2020).

While the Klang Valley only covers a small proportion of Malaysia, its economic activity contributes greatly to the national GDP. Kuala Lumpur's economy originally developed around the processing of locally produced tin and rubber, food products, and traditional handicrafts, but these sectors became less prominent as Malaysia's efforts to develop a more industrialized and export-oriented economy have created new industries in and around Kuala Lumpur. Kuala Lumpur City plays as the premier financial and commercial center in Malaysia. In addition, the employment to population ratio for Kuala Lumpur City is higher if compared with Malaysia as a whole. This shows that Kuala Lumpur City is an important generator of jobs. The three most important subsectors in terms of employment for Kuala Lumpur today are finance, insurance, real estate, and the business services sector (24.2%), followed by wholesale and retail trade, the restaurant and hotel sector, (17.2%) and finally the government services sector (15.1%). This highlights the dominance of producer services supplied by Kuala Lumpur. The government services sector retains a high percentage, but this has changed as many of the federal ministries have shifted to Putrajaya. In the KLC, the three most important employment subsectors are manufacturing (19.8%), followed by finance, insurance, real estate, and business services sector (18.2%), and thirdly personal services (15.2%). This also highlights the growth in the industrial sector in the KLC outside the Kuala Lumpur City center, particularly to the southern part known as Langat River Basin (Er et al., 2013).

5.1.2 Trends and Episodes of Air Pollution

The environmental regulatory agency in Malaysia is the Department of Environment (DoE), under the Ministry of Natural Resources and the Environment. The DoE monitors the country's ambient air quality through a network of 52 continuous monitoring stations, currently through a concessionaire agreement with Alam Sekitar Malaysia Sdn Bhd (ASMA). These monitoring stations are strategically located in both residential and industrial areas to detect any significant change in the air quality which may be harmful to human health and the environment. The National Air Quality Monitoring Network is also supplemented by manual air quality monitoring stations (high-volume air samplers) located at 14 different sites. In addition, MMD monitors selected air quality parameters from a total of 22 stations, mostly located at airports, and some distance away from urban centers across Peninsular Malaysia and East Malaysia.

Parameters monitored by the DoE stations include particulate matter with diameter less than 10 micrometers (PM_{10}) and several heavy metals such as lead are measured once every 6 days. The major gaseous air pollutants monitored include ozone (O_3), sulfur dioxide (SO_2), nitrogen oxides (NO_x), and carbon monoxide (CO). In the case of MMD stations, parameters monitored include rainwater acidity, aerosols (total suspended particulate [TSP] and PM_{10}), and in the Petaling Jaya station, atmospheric O_3 (monitoring of vertical O_3-profile and total column O_3). Most of these air stations are colocated with climatological stations (wind speed, wind direction, temperature, relative humidity, solar radiation, etc.) so that simultaneous and continuous observation of both meteorological and air pollution parameters are carried out. This would ensure that a comprehensive data set comprised of both air quality and meteorological data would be available for assessment of any air pollution event. Furthermore, several other academic institutions are involved in investigating air quality, chemical speciations, and their health concerns.

In general, overall air quality in Malaysia is at a level of good to moderate for most of the time. However, like other countries in the Southeast Asia region, Malaysia also has a fair share of poor air quality episodes. Air pollution in Malaysia is unique by its nature. Seasonal, episodic and transboundary air pollution has been observed for many years. Poor air quality in Malaysia was first associated with the haze of April 1983, which caused severe disruption to daily life. The exact cause of this haze episode was uncertain; however, it has been widely attributed to forest fires from neighboring countries as well as local sources such as agricultural waste burnings, peat soil fires, and fuel combustions from industries and vehicles. Since then, the haze has recurred almost every year, in particular during southwesterly (June to September) and northeasterly (December to March) monsoons with major prolonged episodes recorded in 1991, 1992, 1997, 2003, 2005, and 2013.

Dominick et al. (2012) analyzed the air quality data (CO, NO_2, SO_2, PM_{10} and O_3) obtained from eight DoE monitoring stations across Malaysia during the 2008–2009 period. They reported that average and maximum concentrations recorded during this period were well below the Recommended Malaysian Air Quality Guideline (RMAQG). Main sources of emissions identified were from motor vehicles, aircraft, industries, and the areas of high population density. Azmi et al. (2010) conducted a long-term trend and status of air quality in Klang Valley from 1997 to 2006 using secondary data obtained from the DoE. They found that the concentrations of CO, NO_2, and SO_2 were mainly influenced by heavy traffic while PM_{10} and O_3 were predominantly related to regional tropical factors, such as the influence of biomass burning and of ultraviolet radiation from sunlight with possible local sources (Figure 5.3).

Ahamad et al. (2014) also corroborated the above finding where they reported that the O_3 pattern in the Klang Valley area is strongly influenced by local pollutant emission and dispersion characteristic. Latif et al. (2012) reported that a high surface O_3 concentration is usually observed between January and April, while a low surface O_3 concentration is found between June and August. Analysis of daily variations in surface O_3 and the precursors, NO, NO_2, CO, nonmethane

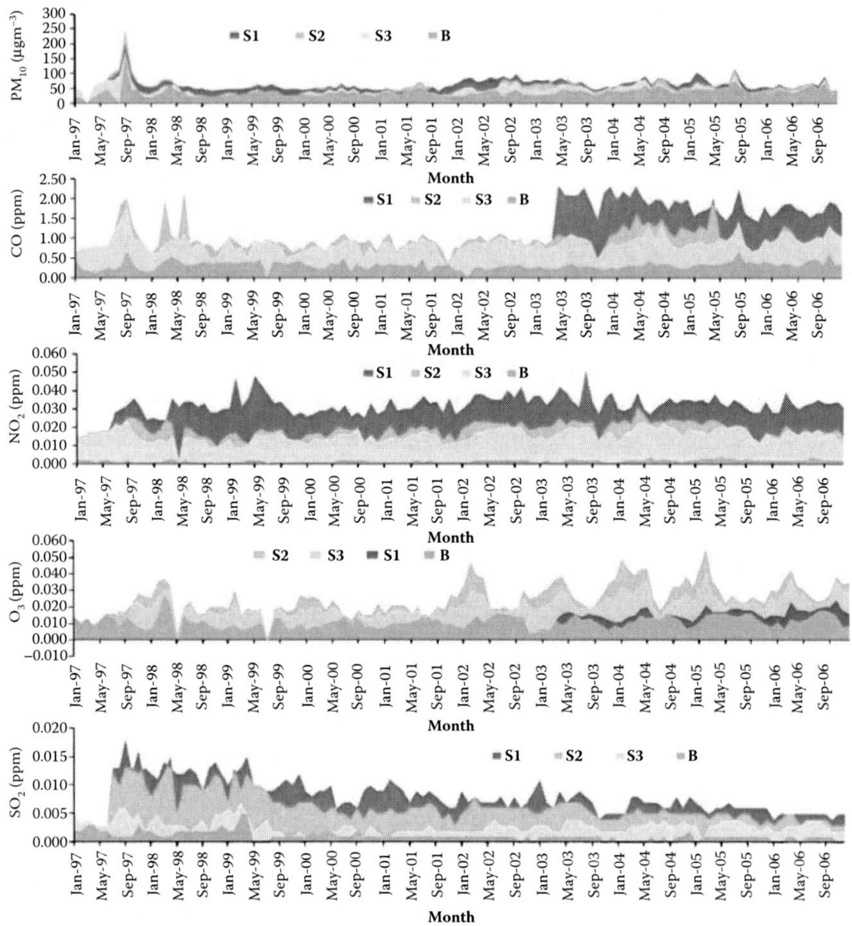

FIGURE 5.3 Monthly trends for air quality in Klang Valley. (From Azmi, S. et al. *Air Quality, Atmosphere and Health,* 3, 53–64, 2010.)

hydrocarbon (NMHC), and Ultraviolet b (UVb), indicated that the surface O_3 photochemistry in this study area exhibits a positive response to the intensity and wavelength in UVb while being influenced by the concentration of NOx, particularly through titration processes.

A study to determine the temporal distribution and chemical characteristics of coarse and fine particulate matter had been conducted in Kuala Terengganu, on the eastern coast of Peninsular Malaysia (Mohd Tahir et al., 2013). They reported that levels of fine (FP: mean = 14.3 ± 6.5) and coarse (CP: mean = 10.4 ± 5.4) particles observed in this city were lower than those reported for the Kuala Lumpur during nonhaze days by Hamzah et al. (2000) (mean FP = 30.9 ± 14.4; mean CP = 24.5 ± 21.1). However, the %FP to PM_{10} was similar, accounting for 58% and 56% for Kuala Terengganu and Kuala Lumpur, respectively.

In addition to the effect of dominant pollution sources, the pattern of local and regional meteorology greatly influences the variability of air pollution in this country. A study by Latif et al. (2014) showed that the meteorological parameters play an important role in air pollution variability. They are closely associated with pollutant concentrations as they influence pollutant dispersion and chemical reactions in photosensitive reactions. Further, precisely the variability of PM_{10} concentration in Malaysia can be decomposed into four dominant modes as described by Juneng et al. (2009): (1) southwest coastal region of the Malaysian Peninsula with the PM_{10} showing a peak concentration during the summer monsoon, that is, when the winds are predominantly southerly or south-westerly, and a minimal concentration during the winter monsoon; (2) the region of western Borneo with the PM_{10} exhibiting a concentration surge in August–September, which is likely to be the result of the northward shift of the intertropical convergence zone (ITCZ) and the subsequent rapid arrival of the rainy season; (3) the northern region of the Malaysian Peninsula with strong bimodality in the PM_{10} concentration (seasonally, this component exhibits two concentration maxima during the late winter and summer monsoons, as well as two minima during the intermonsoon periods); and (4) the northern Borneo region, which exhibits weaker seasonality of the PM_{10} concentration.

The chemical profiles of the reduced visibility were featured by unique characteristics. The concentrations of organic compounds were observed to be greater during the hazy situation than other periods in a year and some of them are suspected to have transported from neighboring areas (Abas and Simoneit, 1996). As an indicator of major organic fraction in atmospheric aerosol, particle-bound polycyclic aromatic hydrocarbons (PAHs) are semivolatile, carcinogenic, and persistent organic pollutants (POPs). A recent study by Jamhari et al. (2014) showed that substantial contributions from traffic emission and a minimal influence from coal combustion and natural gas emissions were observed to the concentrations of PAHs. Surfactants, one of the surface active agents and organic species, are present in the atmospheric aerosol of Kuala Lumpur. The dominant sources of surfactants were motor vehicles, soil/road dust, biomass burning, and sea spray (Wahid et al., 2013).

5.2 HEALTH EFFECTS

Like many developing nations, air pollution is a serious health risk in Malaysia following many years of rapid industrialization and economic growth since independence. To curb air pollution and improve air quality, the Environmental Quality Act (EQA) 1974 and the Environmental Quality (Clean Air) Regulation were introduced in 1978 to control emissions from industry (i.e., power plants and industrial processes) and automobiles, which constitute the major sources of air pollution in urban areas (Afroz et al., 2003). The law also explicitly prohibits open burning, as it exacerbates the problem of haze in the region. Other laws related to air pollution include the Food Act 1983, which controls air pollution from tobacco smoke in public venues; Road Transport Act 1987/Environmental Quality (Control of Emission from Diesel Engine) Regulation 1996, which regulate emission of black smoke by motor vehicles;

the Occupational Safety and Health Act 1994, which sets the limits of airborne concentration for a list of hazardous chemicals in work environments (Mustafa, 2011).

5.2.1 Acute Health Effects of Criteria Air Pollutants

Earliest findings on the acute health effects of air pollutants in Malaysia came from a retrospective time-series study conducted in Klang Valley, the most populated region of Malaysia (Jamal et al., 2004). The study was conducted to quantify short-term health risks associated with daily exposure to the five criteria air pollutants (i.e., PM_{10}, CO, SO_2, NO_2, and O_3). Health endpoints investigated were the respiratory (International Classification of Diseases 10th Revision [ICD10]: J00–J99) and cardiovascular (ICD10: I10–I99) disease-related hospital admissions and nonaccident mortality based on discharge data from two public hospitals (Hospital Kuala Lumpur and Hospital Selayang) and two university hospitals (Hospital Universiti Kebangsaan Malaysia and University Malaya Medical Centre) during the period from January 1, 2000, to December 31, 2003. The study used Poisson general additive model to estimate the relative risk (RR) of daily morbidity and mortality cases for each criteria air pollutant, with adjustment to temporal component, maximum temperature, and rainfall via smoothing function, and day-of-the-week effect via dummy coding. Delayed effects of temperature and rainfall up to the previous 3 days were also adjusted. Single-day effect up to 3 preceding days (lag 3) and cumulative effect using 3- and 5-day moving averages were computed for the first, second (median), and third quartile increments in the concentration of air pollutant. The study reported significant effects of PM_{10}, NO_2, SO_2, CO, and O_3 on respiratory and cardiovascular disease-related hospital admissions for adults and children in all four hospitals based on single-pollutant models. Premature mortality due to the short-term exposure to NO_2, SO_2, and O_3 was also noted. A study that investigated the association between short-term exposure to air pollutants and natural mortality in the Klang Valley region reported significant forward displacement of all-cause mortality as a result of exposure to PM_{10} and O_3 (Wan Mahiyuddin et al., 2013). The study found statistically significant effect of PM_{10} at lag 1 (RR = 1.0099, 95% confidence interval [CI] = 1.009–1.0192) and 5-day cumulative effect of O_3 (RR = 1.0215, 95% CI = 1.0013–1.0202). Given the longer study period of 2000–2006 and that the study was based on mortality statistics from the national database, which captured all mortality cases in the population, statistical power was relatively higher than the previous study that based its findings on the discharge data of four hospitals.

5.2.2 Health Effects of Haze Air Pollution

Despite the various regulations, air pollution remains a public health hazard in Malaysia, with numerous studies reporting adverse health effects of exposure to atmospheric pollutants. Early studies were mostly related to the 1997 Southeast Asian haze, which was due to the widespread open biomass burnings to clear agricultural land and forest in Kalimantan and Sumatra. These large-scale forest and plantation fires caused thick smoky haze over a large portion of Southeast Asia, especially Indonesia, Malaysia, and Singapore (Sastry, 2002), as confirmed via satellite images

(Lim and Ooi, 1998) and analysis of local monitoring data that linked this severe haze episode to high concentration of PM_{10} up to 20-fold beyond the limit recommended by the Malaysian Air Quality Guideline (MAQG, 1989), in spite of the relatively low levels of other gaseous pollutants such as CO_2, NO_2, SO_2, and O_3 compared with the normal nonhaze days (Awang et al., 2000; Noor, 1998).

During the 1997 haze, Hospital Kuala Lumpur, a government tertiary referral hospital, recorded a substantial increase in cases of upper respiratory tract infections, conjunctivitis, and asthma, with a 2-day delayed effect for asthma incidences (Awang et al., 2000). A similar acute trend was observed in other major hospitals in Kuala Lumpur, Sarawak, and neighboring Singapore (Brauer and Hisham-Hashim, 1998), leading to the supposition that instead of other gaseous pollutants, the observed adverse health effects were attributable to the short-term exposure of PM_{10}, the predominant air contaminant during haze (Awang et al., 2000; DoE, 2000). A panel study examining the respiratory function of 107 children found statistically significant decreases of lung function in these children measured between the nonhaze period a year earlier and the haze period in 1997 (Hashim et al., 1998). Another study that used a matched control group to compare the pulmonary functions of 16-year-old schoolchildren exposed to different levels of PM_{10} (i.e., 103 $\mu g/m^3$ vs. 47 $\mu g/m^3$ in the control group) documented significant reduction in spirometry parameters among those with higher long-term exposure to PM_{10} (Awang et al., 2000). More recently, the health effects of haze events occurring between 2000 and 2007 in the Klang Valley region were examined using a case-crossover design with time-stratified referent selection (Sahani et al., 2014). Haze events were defined using a cutoff of PM_{10} concentration at 100 $\mu g/m^3$ based on time-series and backward trajectory analyses. The study reported significant 2-day delayed effect of haze event on all-cause mortality among children less than 14 years of age (odds ratio [OR] = 1.41, 95% CI = 1.01–1.99). Effects of haze events on respiratory mortality were immediate (i.e., current-day, lag 0) (OR = 1.19, 95% CI = 1.02–1.40). This immediate effect on respiratory mortality was particularly discernible among elderly males over 60 years old (OR = 1.41, 95% CI = 1.09–1.84).

5.3 EXPOSURES

5.3.1 Measurement of Air Pollutants in Malaysia

The major air pollutants, for example, TSP, PM_{10}, O_3, SO_2, NOx, and CO, are measured by in-situ monitoring instruments in Malaysia. The TSP, PM_{10}, and trace gases along with other meteorology-related variables (e.g., wind speed, wind direction, temperature, relative humidity, solar radiation) are monitored by DoE. The details of the methods and instruments used are shown in Table 5.1.

5.3.2 Exposure Assessment

Distributions of PM_{10}, O_3, SO_2, NOx, CO, volatile organic compounds (VOCs), as well as heavy metals can be interpreted using synoptic scale wind pattern, cluster of trajectory analysis, as well as local wind direction by the Grid Analysis and Display

TABLE 5.1
Lists of Instruments Used by Department of Environment (DoE) Monitoring Air Quality in Malaysia

Variables	Instrument (Teledyne, US)	Measurement Principal	Precision	Detection Limit (DL)
O_3	Analyzer 400A	Chemiluminescence	0.5% (<10s)	0.04 ppm
NO, NO_2, NO_x	Advanced pollution instrumentation (API) 200A	Chemiluminescence	0.5%	0.4 ppb
SO_2	API M100A	Florescence	0.5%	0.4 ppb
CO	API M300	Nondispersive infrared absorption (NDIR)	0.5% (<10s)	0.04 ppm
CH_4	API M4020	Flame ionization detector (FID)	1%	—
NmHC	API M4020	FID	1%	—
TSP	High volume air sampler (HVAS)	—	—	—
PM_{10}	Beta attenuation monitor (BAM) 1020	Met-One beta attenuation	—	<1.0 μg m^{-3} (24 h)

System (GrADS version 2.0.2), Hybrid Single-Particle Lagrangian Integrated Trajectory model (HYSPLIT 4.9) and wind rose/concentration rose, respectively. Local and regional fire hotspots may also be used to explain the effect of biomass or forest fire with Moderate Resolution Imaging Spectroradiometer (MODIS) fire data.

Multivariate receptor models are very useful tools in the studies of source apportionment of pollutants at urban or local scale. The application of receptor model in principle translates the research results to be used into policy. To introduce control measures, it is necessary to know the source information of pollutants in the amendment of the regulations. Thus, the receptor modeling procedures solely involved in prediction of sources and the contribution of the respective source factor. Several models are widely used by the distinguished researchers for the quantitative information of the pollutant sources. The most commonly used models are (1) chemical mass balance model (CMB) (Watson et al., 1990), (2) positive matrix factorization (PMF) (Paatero and Tapper, 1994), (3) UNMIX (Henry, 1987), and (4) principal component analysis coupled with absolute principal component score (PCA/APCS) (Thurston and Spengler, 1985). Among the receptor models, PMF uses weighted least-squares fit and estimates error of the measured data and can impose nonnegativity constraints. Moreover, the prior source information or prior knowledge of pollutants is not necessary, which is the first and foremost advantage of this procedure. Therefore, the PMF is considered to be used in the apportionment of pollutant sources. Two input files as concentrations of variables and uncertainty of the data are needed to proceed with PMF. The detailed of preparing the uncertainty data file has given in the FMF 3.0 Fundamentals and User Guide by USEPA.

Principal component analysis (PCA) is a statistical technique which takes in the form of eigenvector analysis (Khan et al., 2010). Absolute PCA coupled with multiple linear regression (APCS-MLR) is an advanced version as compared with basic PCA-MLR procedure (Khan et al., 2010; Thurston and Spengler, 1985). The major difference between APCS-MLR and PCA-MLR is in the ability to perform zero correction in the APCS-MLR compared with PCA-MLR. The zero correction reduces the negative indices in the factor scores derived from PCA procedure. Else, a comparison is to be made using PMF 3.0, a robust USEPA developed PMF model, and PCA/APCS as these two receptor modelings are capable of extracting robust source information of pollutants at any local scale.

The dataset to be used in the above receptor modeling procedures needs to undergo a series of pretreatment. The measurement or monitoring data always contain noisy or bad data, outliers, missing and value below detection limit. Prior to feeding into the prescribed models, the data variables are to be cleaned up with proper deletion/replacement procedures acknowledged by peer journals.

5.4 RISK ASSESSMENT AND MANAGEMENT

5.4.1 Source Identification of Particulate Matter

Although numerous studies have established the link between atmospheric particulate matter and adverse health effects in Malaysia, for effective control of air particulate pollutant, information on the sources that contribute to the composition of PM_{10} and its toxicity is important. The earliest work on the chemical characterization and source apportionment of particulate matter based on 1-year monitoring in Klang Valley in 1997–1998 reported that fine particle (particulate matter with size less than 2.5 micron [$PM_{2.5}$]) was the dominant particulate pollutant with level during haze days four to five times higher than that during nonhaze days. Elemental composition revealed both natural and anthropogenic contribution with biomass burning as the primary source during episodes of haze (Hamzah et al., 2000). Another study at Kuala Terengganu, a coastal city east of Malaysian Peninsula, from August 2006 to December 2007 reported that measured chemical species accounted for about 54% and 32% of coarse and fine particle on average, respectively, with the remaining possibly consisting of organic and carbonaceous materials (Mohd Tahir et al., 2013). The study also identified soil dust, marine aerosol, vehicle exhaust, secondary aerosol, traffic aerosol (e.g., nonengine combustion such as tire, clutch, and brake wear), and biomass burning (e.g., garden wastes, use of wood as fuel, and use of palm fiber and shell waste as boiler fuel at the many palm oil mills) as the main sources of PM pollution in the region. A recent study at Bangi, a semiurban area, focusing on heavy metals reported Fe as the dominant element in PM_{10}, followed by Zn and Pb, which were all often associated with traffic emissions, both vehicular and non-engine combustion sources. Earth crust and road dust were the main sources of PM_{10} in the region, followed by vehicle emissions (Wahid et al., 2014). These findings on the chemical mixture and sources of PM are important as they allow further epidemiological work to assess the health impact of specific PM components, instead of the general total-mass-based measure, and can provide important feedback to the current monitoring guideline. A study by Afroz et al. (2003) suggests that the air pollution comes mainly from land transportation, industrial emissions, and open burning sources. Among them, land transportation contributes the most to air pollution.

5.4.2 Economic Burden

Estimates of economic loss due to the health impacts of haze have also been reported. A study reported a country-wide incremental cost of MYR 5.02 million (about USD 1.51 million) for treatment of haze-related diseases, including self-medication, and MYR 4.3 million for productivity losses (about USD 1.29 million) as a result of the 1997 haze pollution (Othman and Mohd Shahwahid, 1999). This study included population at risk from all states in the country except the haze-free states such as Kelantan, Terengganu, and Pahang. The estimation took into consideration the various intensities and length of haze within the August–October period. A more recent economic assessment of haze-related illnesses based on the daily inpatients from four major hospitals in Kuala Lumpur and the surrounding area reported an estimated annual loss of USD 91,000 due to acute exposure to transboundary smoke haze pollutions that occurred in 2005, 2006, 2008, and 2009 (Othman et al., 2014). The figure was expected to go up if outpatient treatment, subsequent productivity loss, and shortage of hospital beds were considered.

5.4.3 Air Quality Management in Malaysia

The EQA, the basic framework for environmental management in Malaysia, was enacted in 1974. The Act was officially endorsed by the Government of Malaysia in its Third Malaysia Plan (1981–1985). The main environmental regulatory agency in Malaysia at the federal level is DoE, which is currently part of the Ministry of Natural Resources and the Environment. It was established to administer and enforce EQA of 1974 (Heng and Looi, 2002).

The Malaysian government plays a very important role in addressing the issues of air pollution. The Malaysian government has already foreseen the importance of managing and tackling the air pollution problems in the country, as these issues have been stated in the two, 9th and 10th, Malaysia Plans (Rancangan Malaysia Ke [RMK]). In the 9th Malaysia Plan, a Clean Air Action Plan (CAAP) was developed and implemented to improve air quality (Abdullah et al., 2012). In the RMK10, some initiatives will be introduced to address climate change through the adoption of strategies to protect economic growth and development factors from the impact of climate change as well as mitigation strategies to reduce the emission of greenhouse gases (EPU, 2010).

The Malaysian government established MAQGs, the Air Pollutant Index (API), and the Haze Action Plan to improve air quality. The Malaysian DoE formulate policy development using the RMAQG 1989 as shown in Table 5.2.

Furthermore, Malaysian DoE introduced API in mitigating the effect of air pollutants as it is an indicator for the air quality status at any particular area. API

TABLE 5.2
Malaysia Recommended Ambient Air Quality Guidelines

Pollutants	Averaging	Malaysian Guidelines ppm	Malaysian Guidelines µg m^{-3}
Ozone (O_3)	1 hour	0.10	200
	8 hours	0.06	120
Carbon monoxide (CO)[a]	1 hour	30.0	35
	8 hours	9.0	10
Nitrogen dioxide (NO_2)	1 hour	0.17	320
	24 hours	0.04	10
Sulfur dioxide (SO_2)	1 hour	0.13	350
	24 hours	0.04	105
Particulate matter with diameter less than 10 micrometer (PM_{10})	24 hours	—	150
	12 months	—	50
Total suspended particulate (TSP)	24 hours	—	260
	12 months	—	90
Lead (Pb)	3 months	—	1.5

[a] mg m^{-3}

is calculated based on the five criteria air pollutants, which are SO_2, NO_2, CO, PM_{10}, and ground-level O_3. The API and its health effect threshold are described in Table 5.3.

5.4.4 Clean Air Action Plan

The CAAP is drawn up in line with the "7th Green Strategy" in the National Policy on the Environment. The CAAP presents a set of strategies and indicators that together provide a roadmap to achieve better air quality by reducing the frequency, severity, and duration of poor air quality episodes.

The strategies and measures listed in this plan are aimed at managing air quality through close cooperation between government, private sectors, and nongovernmental organizations (NGOs). The time frame of this CAAP is categorized based on its priorities, that is, short term (immediate or less than 2 years), medium term (2–5 years), and long term (5–10 years). This plan is a living document; as new technologies and approaches become available, they would be incorporated into the plan.

In the implementation of the CAAP, apart from achieving good air quality it also generates co-benefit in terms of reduction in greenhouse gas emissions. Major contributors of greenhouse gas emissions that are addressed in the plan include emissions from motor vehicles and industries, haze due to land and forest fires, and open burning activities.

TABLE 5.3
Air Pollutant Index (API): Health Effect

API	Status	Health Effect	Health Advice
0–50	Good	Low pollution without any bad effect on health.	No restriction for outdoor activities to the public. Maintain healthy lifestyle.
51–100	Moderate	Moderate pollution that does not pose any bad effect on health.	No restriction for outdoor activities to the public. Maintain healthy lifestyle.
101–200	Unhealthy	Worsen the health condition of high-risk people, who are the people with heart and lung complications.	Limited outdoor activities for the high-risk people. Public need to reduce the extreme outdoor activities.
201–300	Very unhealthy	Worsen the health condition and low tolerance of physical exercises to people with heart and lung complications. Affect public health.	Old and high-risk people are advised to stay indoors and reduce physical activities. People with health complications are advised to see doctor.
>300	Hazardous	Hazardous to high-risk people and public health.	Old and high-risk people are prohibited from outdoor activities. Public are advised to prevent from outdoor activities.
>500	Emergency	Hazardous to high-risk people and public health.	Public are advised to follow orders from National Security Council and always follow the announcement in mass media.

5.4.5 Issues and Challenges

Poor air quality is mainly caused by combustion of fossil and other fuels by industries, motor vehicles, and households. Open burning and forest fires also contribute to poor air quality. Pollutants and hazardous matters are either emitted directly or as the result of chemical reactions of emissions such as ground-level O_3.

The main air pollutants are PM_{10}, NO_2, SO_2, CO, O_3, and VOCs that will post a wide range of negative health impacts such as lung and heart malfunctions, bronchitis and asthma. Inadequate urban planning, the establishment of satellite cities and the preference of individual over public transport result in increasing motor vehicle usage which in turn increases the level of air pollution in urban areas. Low quality of fuel and outdated emission standards further exacerbate the problem. The challenge is to move toward environmentally sustainable transport. Environmentally sustainable transport could result in a number of co-benefits such as reduced air pollution, traffic congestion, and oil usage that improves environmental quality and human health.

Industries without adequate control measures, the use of poor quality fuel and the lack of land-use planning, thus allowing heavy polluting industries to be sited in urban dwelling centers, also contribute to poor air quality. Large-scale and uncontrolled fires resulting from open burning of biomass release significant amounts of pollutants into the atmosphere, including fine dusts, CO, carbon dioxide, and so on. Such fires could result in haze episodes, thus affecting public health and the environment.

5.4.6 Strategies and Actions Taken

The strategies and some of the actions taken in CAAP are outlined into five categories as below:

1. Motor vehicles emission reduction
   Motor vehicles are one of the main contributors to the air pollution in the country, particularly in urban areas. Several strategies and actions had been planned and implemented to reduce motor vehicle emission.
2. Industrial emission reduction
   The air pollutants emitted from the industrial sector comes from various sources such as power plants, industrial energy use, and large-scale industries such as iron and steel plants and cement industries. Several strategies and actions had been planned and implemented to reduce industrial emission.
3. Prevention and control of haze due to land/forest fire and open burning activities
   Several strategies to prevent and control haze pollution at local and regional levels had been planned and implemented.
4. Knowledge enhancement
   Knowledge enhancement aims to establish a scientific and progressive society that is capable, advanced, innovative and forward-looking that contributes to the scientific and technological civilization of the future.

   Under this initiative, the DoE has recently engaged a team of experts on various aspects of atmospheric science from universities and other research institutes, NGOs and private sectors to review the existing Malaysian Ambient Air Quality Guideline. As a result, a new Malaysian Ambient Air Quality Standard has been proposed with an inclusion of a new air parameter, particulate matter with size less than 2.5 micron ($PM_{2.5}$) in addition to existing parameters (PM_{10}, CO, SO_2, NO_2, and ground-level O_3) where their values have been reviewed and revised.
5. Public awareness and participation
   Public support could be achieved through well-informed citizens who are aware and fully committed. Several strategies to enhance public awareness and participation, among others, are to enhance education and awareness for specific target groups at different levels, improve air quality dissemination and feedback mechanism, close partnership at community-based levels, NGO and private sectors, corporate social responsibility and outreach programs to address air pollution issues.

5.4.7 Moving Forward: Air Pollution and Low-Carbon Initiatives

Malaysia recognizes the threat of climate change and has implemented the National Policy on Climate Change (2009) and the National Green Technology Policy (2009) in order to adapt the economy to the low-carbon pathway. In line with the national direction to promote climate-resilient sustainable development, numerous strategies to reduce carbon emission have been identified to avert climate change. Implementation of these low-carbon strategies are currently being studied in selected cities. These strategies to reduce carbon can also improve outdoor air quality, leading to various ancillary health benefits. A study to understand the health co-benefits of emission control is currently ongoing at Iskandar, a designated low-carbon development region in the southernmost tip of Peninsular Malaysia which is undergoing tremendous economic growth. By linking climate change mitigation measures to environmental health, the study hopes to provide additional justification for strict emission cut backs and help stakeholders and policy makers to prioritize mitigation actions against the background of finite resources and time.

5.5 CONCLUSIONS

In this chapter, professionals from various backgrounds have collaborated to construct an overview of air pollution and health in Malaysia. Overall air quality in Malaysia is generally at a level of good to moderate for most of the time. However, like other countries in the Southeast Asia region, Malaysia is also experiencing recurrent haze almost every year, in particular during southwesterly and northeasterly monsoons. Numerous studies have established the link between atmospheric particulate matter and adverse health effects in Malaysia. Further research on the sources that contribute to the composition of PM_{10} and haze pollution is necessary to determine in detail the toxicological effects of local air pollution and haze episodes in Southeast Asia. The air pollution and health risk analysis are important in developing comprehensive decision-support tools in air quality management. The Malaysian government plays a very important role in managing the air quality and has established the RMAQ and the clean air and haze action plan. International cooperation aimed at reducing biomass burning, which could lead to significant public health benefits for Malaysia and other Southeast Asian countries, is really needed for the benefit of people in Southeast Asia.

REFERENCES

Abas, M. R. and Simoneit, B. R. T. 1996. Composition of extractable organic matter of air particles from Malaysia: Initial study. *Atmospheric Environment,* 30, 2779–2793.

Abdullah, A. M., Abu Samah, M. A. and Tham, Y. J. 2012. An overview of the air pollution trend in Klang Valley, Malaysia. *Open Environmental Journal,* 6, 13–19.

Afroz, R., Hassan, M. N. and Ibrahim, N. A. 2003. Review of air pollution and health impacts in Malaysia. *Environmental Research,* 92, 71–77.

Ahamad, F., Latif, M. T., Tang, R., Juneng, L., Dominick, D. and Juahir, H. 2014. Variation of surface ozone exceedance around Klang Valley, Malaysia. *Atmospheric Research,* 139, 116–127.

Awang, M. B., Jaafar, A. B., Abdullah, A. M., Ismail, M. B., Hassan, M. N., Abdullah, R., Johan, S. and Noor, H. 2000. Air quality in Malaysia: Impacts, management issues and future challenges. *Respirology,* 5, 183–196.

Azmi, S., Latif, M., Ismail, A., Juneng, L. and Jemain, A. 2010. Trend and status of air quality at three different monitoring stations in the Klang Valley, Malaysia. *Air Quality, Atmosphere and Health,* 3, 53–64.

Brauer, M. and Hisham-Hashim, J. 1998. Peer reviewed: Fires in Indonesia: Crisis and reaction. *Environmental Science & Technology,* 32, 404A–407A.

Er, E. A., Rostam, K., Nor, A. R. M. and Dali, M. M. 2013. Malaysia: Kuala Lumpur. *In:* Shirley, I. and Neill, C. (eds.) *Asian and Pacific Cities: Development Patterns.* Abingdon, Oxon: Taylors and Francis Group.

Department of Environment. 2000. Environmental Quality Report. Malaysia.

Dominick, D., Juahir, H., Latif, M. T., Zain, S. M. and Aris, A. Z. 2012. Spatial assessment of air quality patterns in Malaysia using multivariate analysis. *Atmospheric Environment,* 60, 172–181.

Economic Planning Unit. 2010. 10th Malaysian plan. EPU, Prime Minister's Department, Malaysia. Available from: http://www.epu.gov.my/html/themes/epu/html/RMKE10/img/pdf/en/chapt6.pdf. Date accessed 5 July 2014.

Economic Planning Unit. 2013. The Malaysian economy in figures. EPU, Prime Minister's Department, Malaysia.

Hamzah, M., Rahman, S., Matori, M. and Wood, A. 2000. Characterisation of air particulate matter in Klang Valley by neutron activation analysis technique. Proceedings of the Research and Development Seminar, Malaysian Institute for Nuclear Technology Research, Bangi, Malaysia, October 17–18, 2000.

Hashim, J., Hashim, Z., Jalaludin, J., Lubis, S. and Hashim, R. 1998. Respiratory function of elementary school children exposed to the 1997 Kuala Lumpur haze. *Epidemiology,* 9, S103.

Heng, J. and Looi, L. 2002. Malaysia. *In:* Mottershead, T. (ed.) *Environmental Law and Enforcement in the Asia–Pacific Rim*, pp. 137–203. Sweet & Maxwell Asia, Hong Kong, Singapore, Malaysia.

Henry, R. C. 1987. Current factor analysis receptor models are ill-posed. *Atmospheric Environment (1967),* 21, 1815–1820.

Jamal, H., Pillay, M., Zailina, H., Shamsul, B., Sinha, K., Zaman, H. Z., Khew, S., Mazrura, S., Ambu, S., Rahimah, A. and Ruzita, M. 2004. *A Study of Health Impact & Risk Assessment of Urban Air Pollution in Klang Valley, Malaysia.* Kuala Lumpur: UKM Pakarunding Sdn Bhd.

Jamhari, A. A., Sahani, M., Latif, M. T., Chan, K. M., Tan, H. S., Khan, M. F. and Mohd Tahir, N. 2014. Concentration and source identification of polycyclic aromatic hydrocarbons (PAHs) in PM10 of urban, industrial and semi-urban areas in Malaysia. *Atmospheric Environment,* 86, 16–27.

Juneng, L., Latif, M. T., Tangang, F. T. and Mansor, H. 2009. Spatio-temporal characteristics of PM10 concentration across Malaysia. *Atmospheric Environment,* 43, 4584–4594.

Katiman, R. 1997. Industrial expansion, employment changes and urbanization in the peri-urban areas of Klang-Langat Valley, Malaysia. *Asian Profile,* 25, 303–315.

Khan, M. F., Hirano, K. and Masunaga, S. 2010. Quantifying the sources of hazardous elements of suspended particulate matter aerosol collected in Yokohama, Japan. *Atmospheric Environment,* 44, 2646–2657.

KL. 2020. Kuala Lumpur structure plan. Available from www.dbkl.gov.my/pskl2020/english/international and_national_context_of_growth/index.htm. Date accessed 19 October 2014.

Latif, M. T., Dominick, D., Ahamad, F., Khan, M. F., Juneng, L., Hamzah, F. M. and Nadzir, M. S. M. 2014. Long term assessment of air quality from a background station on the Malaysian Peninsula. *Science of the Total Environment,* 482–483, 336–348.

Latif, M. T., Huey, L. S. and Juneng, L. 2012. Variations of surface ozone concentration across the Klang Valley, Malaysia. *Atmospheric Environment,* 61, 434–445.

Lim, J. and Ooi, S. 1998. Effects of biomass burning in Southeast Asia on meteorological conditions in Malaysia. National Symposium on the Impact of Haze, Universiti Putra Malaysia.

Malaysia Meteorological Department. 2015a. General climate information. MMD. Available from http://en.wikipedia.org/wiki/Geography_of_Malaysia. Date accessed 17 February 2015.

Malaysia Meteorological Department. 2015b. General climate of Malaysia. MMD. Available from http://www.met.gov.my/index.php?option=com_content&task=view&id=75&Itemid=1089). Date accessed 17 February 2015.

Mohd Tahir, N., Suratman, S., Fong, F. T., Hamzah, M. S. and Latif, M. T. 2013. Temporal distribution and chemical characterization of atmospheric particulate matter in the Eastern Coast of Peninsular Malaysia. *Aerosol and Air Quality Research,* 13, 584–595.

Mustafa, M. 2011. Environmental law in Malaysia, the Netherlands: Kluwer Law International.

Noor, H. 1998. Haze and health: Malaysian experience. Workshop on the Impacts of Transboundary Haze Pollution of the 1997/98 Fire Episodes on Health, Kuala Lumpur, Malaysia.

Othman, J. and Mohd Shahwahid, H. O. 1999. Cost of trans-boundary haze externalities. *Journal Ekonomi Malaysia,* 33, 3–19.

Othman, J., Sahani, M., Mahmud, M. and Sheikh Ahmad, M. K. 2014. Transboundary smoke haze pollution in Malaysia: Inpatient health impacts and economic valuation. *Environmental Pollution,* 189, 194–201.

Paatero, P. and Tapper, U. 1994. Positive matrix factorization: A non-negative factor model with optimal utilization of error estimates of data values. *Environmetrics,* 5, 111–126.

Sahani, M., Zainon, N. A., Wan Mahiyuddin, W. R., Latif, M. T., Hod, R., Khan, M. F., Tahir, N. M. and Chan, C.-C. 2014. A case-crossover analysis of forest fire haze events and mortality in Malaysia. *Atmospheric Environment,* 96, 257–265.

Sastry, N. 2002. Forest fires, air pollution, and mortality in Southeast Asia. *Demography,* 39, 1–23.

Thurston, G. D. and Spengler, J. D. 1985. A quantitative assessment of source contributions to inhalable particulate matter pollution in metropolitan Boston. *Atmospheric Environment (1967),* 19, 9–25.

Wahid, N. B. A., Latif, M. T., Suan, L. S., Dominick, D., Sahani, M., Jaafar, S. A. and Mohd Tahir, N. 2014. Source identification of particulate matter in a semi-urban area of Malaysia using multivariate techniques. *Bulletin of Environmental Contamination and Toxicology,* 92, 317–322.

Wahid, N. B. A., Latif, M. T. and Suratman, S. 2013. Composition and source apportionment of surfactants in atmospheric aerosols of urban and semi-urban areas in Malaysia. *Chemosphere,* 91, 1508–1516.

Wan Mahiyuddin, W. R., Sahani, M., Aripin, R., Latif, M. T., Thach, T.-Q. and Wong, C.-M. 2013. Short-term effects of daily air pollution on mortality. *Atmospheric Environment,* 65, 69–79.

Watson, J. G., Robinson, N. F., Chow, J. C., Henry, R. C., Kim, B. M., Pace, T. G., Meyer, E. L. and Nguyen, Q. 1990. The USEPA/DRI chemical mass balance receptor model, CMB 7.0. *Environmental Software,* 5, 38–49.

6 Air Pollution and Health in the Republic of Korea

Dong-Chun Shin

CONTENTS

6.1 BACKGROUND

6.1.1 Demographics of the Republic of Korea

The Republic of Korea (ROK) is a country in the southern part of the Korean peninsula, which stretches south from the northeast coast of Asia. It shares oversea borders with China to the west and Japan to the east. The size of land is 99,392 km^2, and the number of residents is approximately 50 million, including approximately 10 million people in Seoul, the capital. There are six metropolitan cities (Incheon, Daejeon, Daegu, Ulsan, Gwangju, and Busan), and each city has over 1 million residents.

The ROK is located in the midlatitude temperate climate zone (33°–43° north latitude, 124°–132° east longitude) with a humid continental and subtropical climate, and the spring, summer, autumn, and winter seasons are distinct. Except in the mountains and islands, the annual average temperature is 10°C–15°C, and the monthly average ranges from −6°C to 3°C in January to 23°C–26°C in August. Throughout the year, the amount of rainfall is 1200–1500 mm in the Northern provinces and 1000–1800 mm in the Southern provinces. Seasonally, 50%–60% of the annual precipitation occurs in summer. The relative humidity is 60%–75% throughout the year, 70%–85% in July and August, and 50%–70% in March and April.

6.1.2 Economic Growth and Increase in Traffic and Air Pollution

The ROK has experienced dramatically accelerated urbanization resulting from rapid economic development and increased industrialization. From 1960 to 1990, the population increased by nearly 17 million, and the gross domestic product (GDP) increased more than 130 times during the same period, from 200 million dollars to 263.7 billion dollars (Korea National Statistical Office, 2008). During the same period, the population in urban areas was only 27.7% in 1960 but increased to 73.8% in 1990 (United Nations, 2012).

In terms of energy use, the ROK used to mostly depend on coal and wood for fuel in the 1960s, but oil has become the largest energy source since the 1970s. In the 1980s, the government adopted air pollution reduction policies including the supply of low-sulfur fuel, lead-free fuel, and liquefied petroleum gas, which might contribute to a decrease in the levels of several major air pollutants such as sulfur dioxide (SO_2), carbon monoxide (CO), and lead (Kim, 2013). However, there has been a drastic increase in the number of vehicles from 130,000 in 1970 to 12.1 million in 2000 (Korea National Statistical Office, 2008), and traffic-related air pollution has been an issue in urban areas.

Recently, China has been experiencing extremely rapid economic development that has imposed significant pressure on the environment (Asia Pacific Energy Research Centre, 2004). Air pollutants emitted in China and carried by the westerly winds appeared to contribute to 20%–40% of sulfur oxides (SOx) (National Institute of Environmental Research [NIER], 2009) and up to 65% of nitrogen oxides (NOx) (NIER, 2012) in the ROK. Furthermore, because meteorological factors are related to the generation of secondary pollutants such as O_3, the influence of climate change on air pollutants cannot be ignored. Therefore, although the ROK has endeavored to improve air quality for several decades, air pollution still remains problematic and deserves more attention.

6.2 HEALTH EFFECTS

6.2.1 Mortality

The association of air pollution with mortality in the ROK has been investigated since the late 1990s, focusing on acute effects through time-series or time-stratified case-crossover studies. A study of the former type suggested that an increase of 100 $\mu g/m^3$ in total suspended particulates (TSP) was related to a 3% increase in daily mortality in Ulsan, an industrialized city, from 1991 to 1994 (Lee et al., 1998). Kwon and Cho (1999) found that the risk of daily mortality due to air pollution increased in Seoul from 1991 to 1995. They showed that air pollutants such as O_3, nitrogen dioxide, TSP, and SO_2 were associated with the risk of mortality, and a remarkable association was shown for O_3 (6% change; 95% confidence interval [CI], 2%–10% per 100 ppb).

In 1994, the Ministry of Environment started to monitor, in addition to TSP, particulate matter ≤10 μm in aerodynamic diameter (PM_{10}), and since then, researchers have reported an association between PM_{10} and daily mortality. We analyzed death statistics and air pollution data from an automatic monitoring system (1999–2008) in Seoul with a time-stratified case-crossover study design (Shin, 2010). The increased risks for all-cause mortality related to PM_{10} were statistically significant in the cumulative lag2 model, averaging concentrations from the same day to the subsequent 2 days before death (odds ratio [OR], 1.005; 95% CI, 1.003–1.007 per 22 μg/m^3). Regarding cause-specific mortality, both cardiovascular and cerebrovascular deaths were significantly associated with PM_{10} exposure in the cumulative lag2 models. The highest risks were observed in the cumulative lag1 model for deaths from myocardial infarction (OR, 1.011; 95% CI, 1.001–1.021). Ha et al. (2011) conducted a further study in the same manner, expanded to seven major cities (Seoul, Incheon, Daejeon, Daegu, Ulsan, Gwangju, and Busan). They analyzed deaths of people ≥40 years old (n = 160,273) from 2002 to 2008, and the results from each city were combined using a random-effects meta-regression model. An interquartile range (IQR) (39.05 μg/m^3) increase in PM_{10} was associated with the risk of death from cardiovascular and cerebrovascular disease (0.8% change; 95% CI, 0.1%–1.6%).

Ample evidence has informed us about the relationships between air pollution and mortality, but in terms of public health implications, it is imperative to identify for whom and when the risk is higher for death due to air pollution. The study of Ha et al. (2011) also analyzed the data after stratification by age (40–64 and ≥65 years of age) and found a significant risk for cardiovascular and cerebrovascular mortality only in the elderly. Another study using the data from 2000 to 2006 in Seoul also showed a significantly increased risk for all-cause mortality in the ≥66-year-old group, but the increased risk was not significant in the <66-year-old group (Yi et al., 2010). These results underscore the need to treat the elderly as susceptible to the effects of air pollution. In addition, Yi et al. (2010) focused on a seasonal effect in the association between PM_{10} concentrations and mortality, suggesting significantly increased risks for nonaccidental mortality in summer and for cardiovascular mortality in autumn. To explain this seasonal difference, they suggested that individual behaviors related to exposure to air pollutants might affect the seasonality of the association.

6.2.2 Respiratory System

There has been a long-standing concern about the health effects of air pollution on the respiratory system. In the 1960s, the first study in Korea reported that daily visits of patients were associated with ambient SO_2 and CO concentration in Seoul (Chung, 1969). In the 2000s, with more information on air pollutants, epidemiologic studies on the effects of air pollution on the respiratory system have primarily focused on acute effects using medical care utilization and pulmonary function data. A panel study assessed the influence of PM_{10} on normal children's lung function (Kim et al., 2005). The researchers found that the forced expiratory volume in 1 second (FEV1) and forced vital capacity (FVC) were lower in relatively high PM_{10} conditions. Another study found that O_3 was associated with lung function using interpolation methods (Son et al., 2010). An 11 ppb increase in O_3 was associated with a 6.1% (95% CI, 5.0%–7.3%) decrease in FVC and a 0.50% (95% CI, 0.03%–0.96%) decrease in FEV1.

Among respiratory diseases, asthma has been of main interest, especially in children. Lee et al. (2002) analyzed daily admission for asthma, using the Medical Insurance Corporation's reports from all hospitals of only patients younger than 15 years of age who lived in Seoul from 1997 to 1999. They conducted a time-series study of 6436 asthma-related hospital admissions. The relative risk (RR) of hospitalization for asthma was 1.07 (95% CI, 1.04–1.11) for PM_{10} (IQR, 40.4 μg/m^3); SO_2, nitrogen dioxide (NO_2), O_3, and CO were also associated with hospitalization for asthma. The importance of identifying a susceptible group was also highlighted in a study of Park et al. (2013), who conducted a time-series study from 1999 to 2003 in seven metropolitan cities. Using 15- to 64-year-old subjects as the reference, the relative rate of asthma admissions with a 10 μg/m^3 increase of PM_{10} is 1.5% (95% CI, 0.1%–2.8%) lower for children and 1.3% (95% CI, 0.7%–1.9%) higher for the elderly; the relative rate with a 1 ppm increase of CO is 1.9% (95% CI, 0.3%–3.8%) lower for children; and the relative rate with a 1 ppb increase of NO_2 (1 ppb) is 0.5% (95% CI, 0.3%–0.7%) higher for the elderly. Recently, our group conducted a time-series study to evaluate the risk of emergency department (ED) visits for asthma exacerbation related to air pollutants. The study evaluated ED visits from 2005 to 2009 in Seoul. During the study period, 27,896 asthma attack cases were observed. The risk of an ED visit for an asthma attack increased by 5.3% (95% CI, 3.7%–7.0%) with an increase of O_3 (standard deviation, 0.01 ppm) on the same day.

Using biological monitoring data, Kim et al. (2005) suggested a relationship of volatile organic compounds (VOCs) and polycyclic aromatic hydrocarbons (PAHs) with the occurrence of asthma in children. They assessed their exposure levels to VOCs by measuring urinary concentrations of hippuric acid and muconic acid (for VOCs) and 1-OH-pyrene and 2-naphthol (for PAHs) in 30 children with asthma (cases) and 30 children without asthma (controls). The mean concentration of muconic acid and the mean level of urinary 1-OH-pyrene were higher in the asthma group than in the control group. They suggested that VOCs and PAHs have some role in asthma.

Another group focused on relationship between long-term exposure and lung cancer. Using random-intercept Poisson regression and an empirical Bayesian method, they calculated the standardized incidence ratio and standardized mortality ratio. The estimated percent increases in the rate of female lung cancer incidence and mortality were 65% and 27%, respectively, at the highest PM_{10} category ($\geq$70 μg/m^3) compared with the reference category (<50 μg/m^3).

Despite a high economic status of the ROK, tuberculosis (TB) remains an important health problem in the country. According to a retrospective cohort study of the population of Seoul (Hwang et al., 2014), from 2002 to 2006, 41,185 TB cases were reported to the Korean Institute of Tuberculosis. An IQR increase (0.3 ppb) in outdoor SO_2 concentration was associated with a 7% increase in TB incidence (RR, 1.07; 95% CI, 1.03–1.12).

Based on these studies, we can conclude that air pollution affects the health of the respiratory system. Air pollution is related to decreases in lung function, asthma and its exacerbation, lung cancer, and even TB.

6.2.3 Cardiovascular System

There is increasing evidence about the adverse cardiovascular or cerebrovascular effects of air pollution in Korea. Several epidemiologic studies have shown the potential outcomes, including increased daily mortality, hospital admissions, and heart-rate variability (HRV).

An early study was conducted about susceptibility in patients with congestive heart failure mediated by cardiovascular mechanisms (Kwon et al., 2001). The effect of air pollution on daily mortality of two study populations, a general population and congestive heart failure patients among residents of Seoul, was analyzed by general additive Poisson regression in a case-crossover study from 1994 to 1998. The estimated effects appeared larger among the congestive heart failure patients than among the general population (approximately 2.5–4.1 times higher depending on the pollutants). The finding that patients with congestive heart failure were more susceptible to air pollution strengthens the evidence that an important mechanism of the air pollution effect involves the cardiovascular system.

Another study also reported the effects of ambient air pollution on hospital admissions for ischemic cardiovascular diseases within an elderly population in Seoul (Lee et al., 2003). The RRs of hospitalization associated with an IQR increase in pollution concentrations were estimated by using a generalized additive Poisson model in a time-series analysis. The estimated effects of hospitalization were 1.05 (95% CI, 1.01–1.10) for PM_{10}, 1.10 (95% CI, 1.05–1.15) for O_3, 1.08 (95% CI, 1.03–1.14) for NO_2, and 1.07 (95% CI, 1.01–1.13) for CO. The finding that hospital admission for ischemic heart diseases was associated significantly with ambient air pollutants may provide the insights that the elderly appear to be at particular risk from the effects of air pollution, at pollutant levels lower than the standards adopted by most governments.

Min et al. (2008) studied the effects of ambient air pollutants on cardiac autonomic function by measuring HRV in 1349 community residents in Korea (596 men and 753 women). Linear regression analyses were carried out to evaluate the association over 72 hours, and the parameters of HRV indices were presented as the percent change. The findings that the exposures to PM_{10}, SO_2, and NO_2 were significantly associated with reduced HRV indices suggest that air pollutants stimulate the autonomic nervous system, provoke an imbalance in cardiac autonomic control, and may lead to pathological consequences particularly in high-risk patients and susceptible subjects.

Similarly, a few studies were also conducted to assess the association between ambient air pollution and cerebrovascular disease such as acute stroke. Hong et al. (2002) investigated the association between acute stroke mortality and air pollution over a 7-year period from 1991 to 1997 in Seoul. A time-series analysis with a generalized additive model was used to examine the effects of air pollutants on ischemic and hemorrhagic stroke deaths, and significantly increased RRs were found for ischemic stroke mortality for TSP and SO_2. However, as dramatic improvements in the treatment for acute fatal strokes have resulted in a decrement in stroke mortality, it is now much more important to consider the incidence of stroke to evaluate the disease burden of acute stroke.

More recently, our group conducted another study to examine the short-term effect of ambient air pollutants on the incidence of acute stroke. A time-series design using generalized additive models with Poisson regression was applied to evaluate ED visits in Seoul from 2005 to 2009. The RRs of emergency visits for stroke were 1.016 (95% CI, 1.011–1.023) per 12.04 ppb increment of NO_2 and 1.014 (95% CI, 1.007–1.020) per 0.24 ppm increment of CO, and our results suggest that ambient air pollution increases the risk of cardiovascular and cerebrovascular disease in Korea.

Air pollution contributes to the consistent increased risk for cardiovascular or cerebrovascular events in relation to both short- and long-term exposure by several plausible pathways. Oxidative stress as a critically important cause and consequence of PM-mediated cardiovascular effects has a sound experimental basis, and some of the general pathways including systemic inflammation, autonomic nervous system imbalance, and PM or its constituents reaching the systemic circulation can be capable of eliciting cardiovascular events (Brook et al., 2010). Recently, our group investigated the arrhythmogenic mechanism of PM via oxidative stress and calcium calmodulin kinase II activation that might help to explain the relationship between air pollution and increased arrhythmia (Kim et al., 2012). Nevertheless, there is not enough research on air pollution and cardiovascular diseases in Korea, and more investigations are required because the underlying mechanisms involved are not yet fully understood.

6.2.4 Reproductive System

Neonates, infants, and children are considered susceptible to air pollution because their respiratory and immune systems are immature (Braga et al., 2001). During pregnancy, gaseous pollutants such as CO can be transmitted to the fetus via the placenta (Longo, 1977), and systemic inflammation induced by particulates might increase blood viscosity (Peters et al., 1997). Collectively, these mechanisms can lead to low oxygen delivery in the fetus and consequently low birth weight and infant mortality.

In the ROK, the relationship between air pollution and reproductive outcomes was first explored by Ha et al. (2001), who hypothesized that air pollution exposure during pregnancy is associated with low birth weight in full-term births. In generalized additive logistic regression models, CO, NO_2, SO_2, and TSP exposure in the first trimester of pregnancy increased the risk for low birth weight. This research group conducted a further study to find a specific month of pregnancy during which exposure resulted in a high risk for low birth weight (Lee et al., 2003). They suggested that low birth weight was associated with CO, NO_2, SO_2, and PM_{10} exposure in months 2–5, 3–5, 3–5, and 2–4 during pregnancy, respectively.

Another stream of research on reproductive outcome examines infant mortality. To investigate the acute effect, Ha et al. (2003) compared mortality among infants, a 2- to 64-year-old group, and a ≥65-year-old group. Among all age groups, infants were at the highest risk for all-cause mortality (RR, 1.142; 95% CI, 1.096–1.190; per 42.9 μg/m^3) and respiratory mortality (RR, 2.018; 95% CI, 1.784–2.283; per 42.9 μg/m^3) related to PM_{10} exposure. In terms of a long-term effect, a study focused on infant mortality due to gestational and lifetime exposure to particulate matter using survival analysis (Son et al., 2011). The authors found positive associations of gestational exposure to particulate matter with all-cause and respiratory mortality in infants.

6.3 EXPOSURES

6.3.1 Air Pollution Monitoring Network

The national air pollution monitoring network examines the status of the ambient air pollution nationwide and contributes to the advancement of air quality control policies in the ROK (Table 6.1). The national air pollution monitoring network provides information on air pollution to the public via the National Ambient Air Monitoring System (NAMIS), and the Airkorea (Real-time Ambient Air Quality Monitoring System). The NAMIS system collects, selects, and creates the statistics on the data monitored by the ambient air monitoring network nationwide and provides these data to administrative organizations on a government- and a self-governing level. It shares the information on the ambient pollution and utilizes it as the basic data to set the ambient air conservation

TABLE 6.1
Types of Ambient Air Monitoring Networks in Korea (Based on December 2012 Data)

Classification		Measured Items	Purpose of Installation	No. of Sites
General networks	Urban air quality	SO_2, nitrogen oxides (NOx), CO, O_3, particulate matter ≤10 μm in aerodynamic diameter (PM_{10}), wind direction, wind speed, temperature, humidity (continuous/hourly)	Identify average air quality concentration of urban area to determine if environmental standard is achieved	250
	Suburban air quality	SO_2, NOx, CO, O_3, PM_{10}, wind direction, wind speed, temperature, humidity (continuous/hourly)	Identify background concentration of suburb area surrounding a city	19

(Continued)

TABLE 6.1 (*Continued*)
Types of Ambient Air Monitoring Networks in Korea (Based on December 2012 Data)

Classification		Measured Items	Purpose of Installation	No. of Sites
	National background	SO_2, NOx, CO, O_3, PM_{10}, particulate matter ≤2.5 µm in aerodynamic diameter ($PM_{2.5}$), wind direction, wind speed, temperature, humidity (continuous/hourly)	Identify national background concentration, and identify the status of inflow of pollutants from overseas, outflow of pollutants to overseas, and so on	3
	Roadside	SO_2, NOx, CO, PM_{10}, hydrocarbon (HC), wind direction, wind speed, temperature, humidity(continuous/ hourly)	Identify air quality of the roadside with heavy road traffic and pedestrian traffic	38
Special networks	Acid deposition	pH, Cl^-, NO_3^-, SO_4^{2-}, NH^{4+}, Na^+, K^+, Ca^{2+}, Mg^{2+}, electrical conductivity, precipitation, and so on (precipitation)	Identify dry deposition of pollutants from atmosphere and wet deposition of pollutants from rainfall, snowfall, and so on	40
	Photochemical assessment monitoring stations	O_3, NOx, PM_{10}, $PM_{2.5}$, volatile organic compounds (VOCs) (55 types including ethane), carbonyl, insolation, amount of ultraviolet rays, temperature, humidity, wind direction, wind speed, precipitation (continuous/hourly)	Identify concentration of VOCs, the sources of O_3 pollution in urban areas, and closely examine the O_3 pollution phenomenon and utilize as basic data for O_3 forecast	27
	Visibility	Visibility, $PM_{2.5}$ (continuous/hourly)	Measure the visibility of urban atmosphere and also identify the felt contamination	18
	$PM_{2.5}$ monitoring	$PM_{2.5}$, elemental carbon, organic carbon (daily)	Identify the levels of $PM_{2.5}$ and emission sources	11

(*Continued*)

TABLE 6.1 (*Continued*)
Types of Ambient Air Monitoring Networks in Korea (Based on December 2012 Data)

Classification		Measured Items	Purpose of Installation	No. of Sites
	Hazardous air pollutants	VOCs (13 types including benzene, toluene), polycyclic aromatic hydrocarbons (PAHs) (seven types including benzo(a)anthracene) (monthly)	Identify the contamination state by specific hazardous air pollutants in urban areas or nearby industrial complexes	31
	Heavy metals	During regular measurement: Pb, Cd, Cr, Cu, Mn, Fe, Ni; during yellow dust measurement: Pb, Cd, Cr, Cu, Mn, Fe, Ni, Al, Ca, Mg (5 times/month)	Identify the contamination state by heavy metals in urban areas or nearby industrial complexes	52
	Global climate change	CO_2, chlorofluorocarbons (CFCs), N_2O, CH_4 (continuous/hourly)	Identify concentration of global warming substances and O_3-depleting substances in the air	1

policy. The data and information on the ambient air pollution gathered by the ambient air quality monitoring network are provided in real time on the Airkorea website for the public. Airkorea describes ambient air quality based on the health risk of air pollution (Korea Environment Corporation, 2012; Airkorea, 2012).

6.3.2 Current Status of Air Pollution

During the last 10 years in the ROK, the annual averages of SO_2, CO, and NO_2 concentrations in ambient air were approximately 0.05, 0.1, and 0.02 ppm, respectively (Figure 6.1). Lead is in a declining trend with an annual level of 0.05. In contrast, the annual average of O_3 concentration was 0.025 ppm in 2012 and slightly rising every year (Korea NIER, 2013). Especially, the PM_{10} concentration in the Seoul metropolitan area in the early 2000s was high, at approximately 70 μg/m^3 (Figure 6.2).

In order to protect the health of Korean citizens, the Korean government established and promoted a basic plan for air quality management of the capital area, such as installation of emission reduction devices on diesel engine vehicles and low-pollution vehicle supply projects (393,837 vehicles), total air pollution load management systems at places of business (304 places), and low NOx burner installation projects (2931 vehicles). In the meantime, the PM_{10} concentration has been improving every year; the PM_{10} concentration in the capital area was reduced

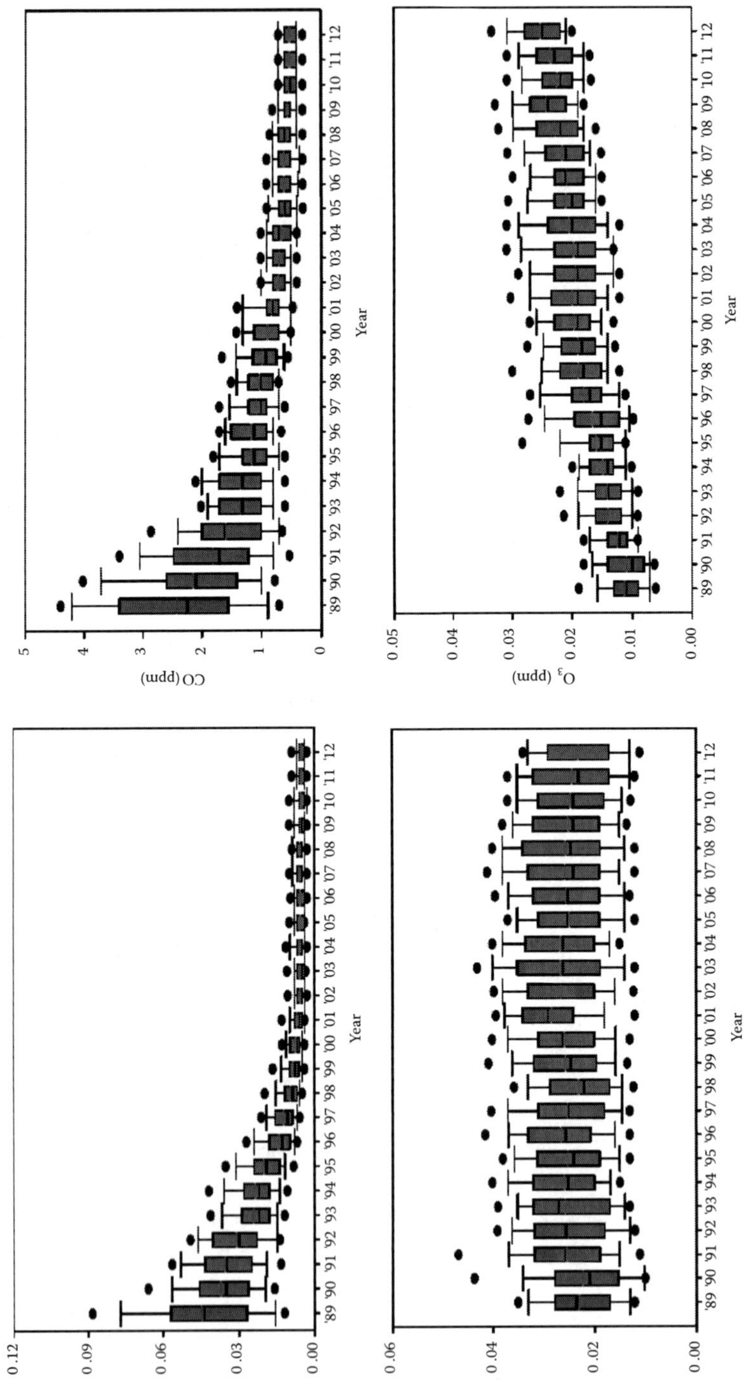

FIGURE 6.1 Annual trend of gaseous air pollutants in the Republic of Korea.

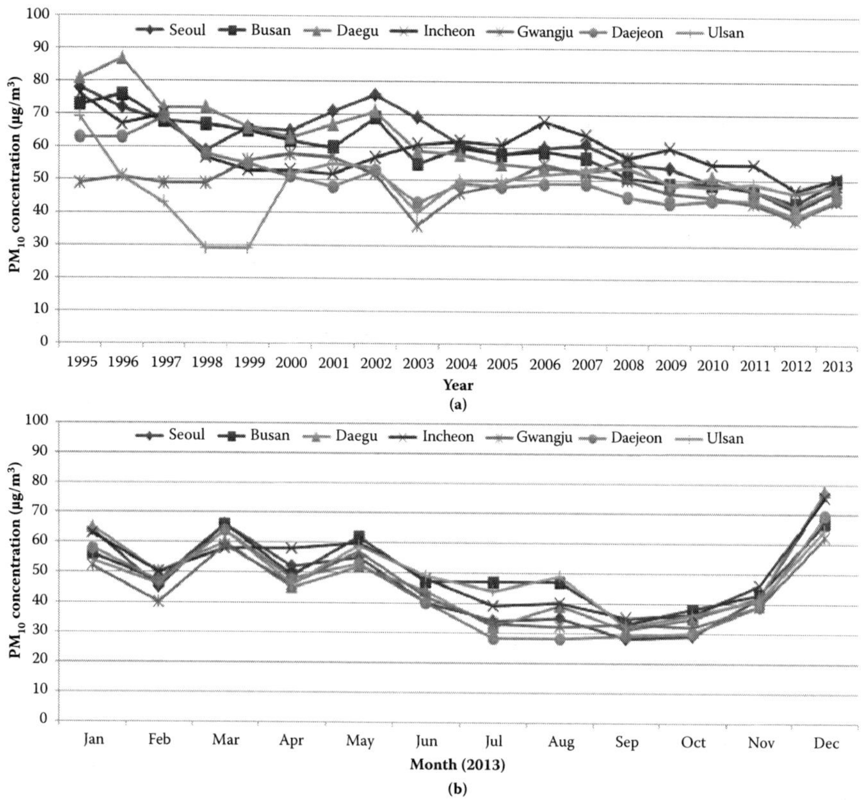

FIGURE 6.2 (a) Annual and (b) 2013 monthly trends of PM_{10} in the Republic of Korea.

to 51 µg/m^3 in 2011, and the number of days that environmental requirements were satisfied was increased to 205 days in 2011 (ECOREA, 2013). The analysis of long-term trends of PM_{10} levels indicates a weak but consistent decline in concentrations in most cities with relative average annual reductions between 0.4% and 2.8% per year (Sharma et al., 2014).

According to monitoring conducted by the NIER of Korea at four air pollution monitoring stations across the country during 2011, the density of the particulate matter ≤2.5 µm in aerodynamic diameter ($PM_{2.5}$) on Baekryeong Island was 23.9 µg/m^3, whereas it was 29.3, 30.9, and 32.4 µg/m^3 in the Seoul metropolitan, southern, and central regions, respectively. Notably, the density of $PM_{2.5}$ in Korea is twice as high as that of major cities in the United States and exceeds 25 µg/m^3, the atmosphere environmental standard that will be applied in 2015 (Energy Korea, 2012).

The concentrations of metals in ambient air are generally lower than the national air standard or World Health Organization air quality guidelines. The annual averages of Pb, Cd, Cr, Cu, Mn, Fe, Ni, and As in 2012 were 0.05, 0.001, 0.01, 0.04, 0.07, 1.3, 0.006, and 0.005 µg/m^3, respectively (Figure 6.3). The concentration of each VOC ranged from nondetectable to tens of parts per billion, with toluene, ethylbenzene, and xylenes being mainly detected. Annual levels of benzene were 0.12–2.56

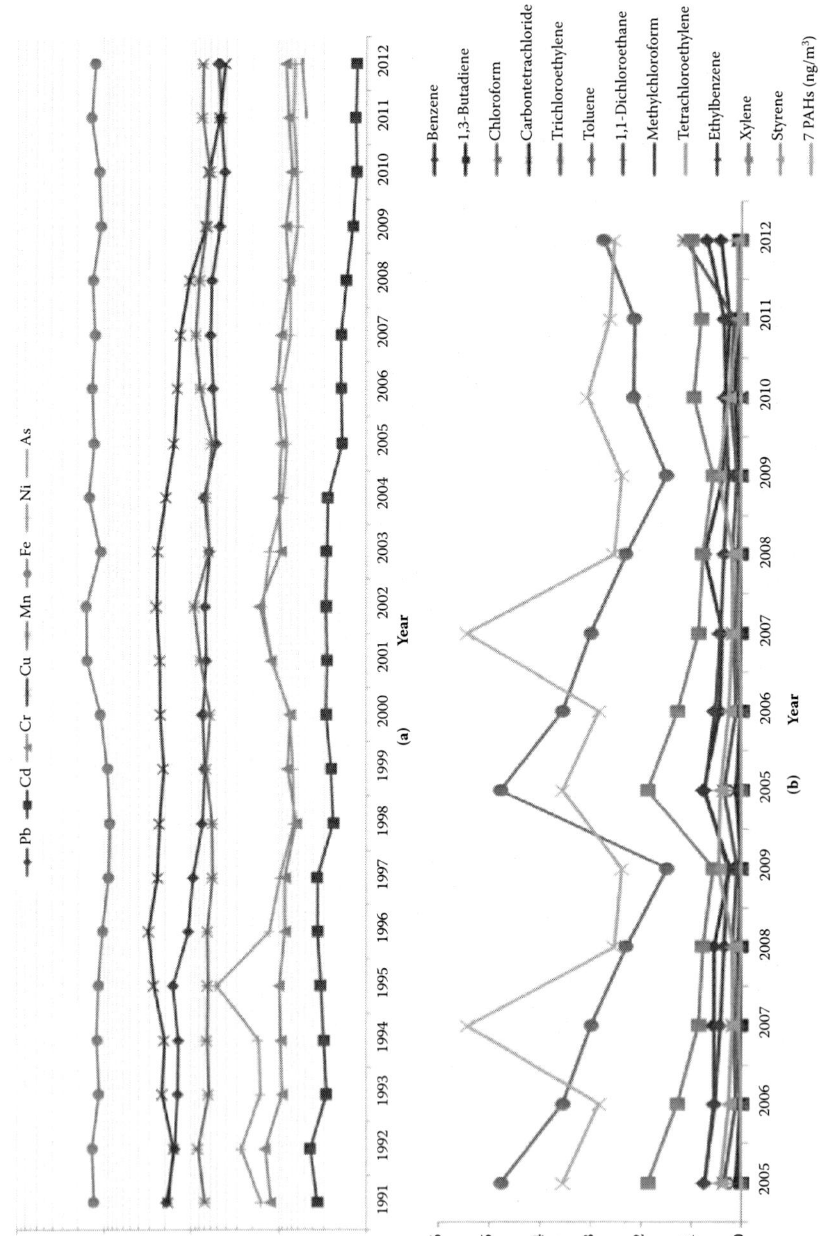

FIGURE 6.3 Annual trends of (a) metals and (b) hazardous air pollutants in the Republic of Korea.

ppb, and the maximum level at an industrial area exceeded the annual ambient air standard of Korea (1.5 ppb) (Korea NIER, 2013).

Recently, the ROK and Japan have sounded the alarm regarding potentially hazardous air pollution from northern China, which is expected to worsen this winter. On a day in February 2013, the average hourly density of PM_{10} during the day exceeded 100 μg/m^3 in Seoul (Airkorea, 2014). The fine dust that is smaller than 2.5 μm is much more harmful to the human body than PM_{10}. Therefore, a recent study of air quality evaluated $PM_{2.5}$ while focusing on observation-based sources, formation processes, and long-range transport through six intensive air quality monitoring stations nationwide. International joint programs, such as the Korea–China–Japan LTP project (studying long-range transboundary air pollutants in Northeast Asia) and a Korea–China expert network on Asian dust prevention, are also being planned to carry out research on air quality improvement as well as research on O_3 and aerosols.

Recent technological advances in air quality measurement methods now make it possible and practical to monitor air pollution. Researchers now use X-ray fluorescence spectrometry to measure air pollution metal content, ion chromatography to identify other chemicals, and other methods to assess organic and elemental carbon levels. Nevertheless, the current measurement technologies do not take into account the relative toxicity of the various components in order to provide an explanation of how air pollution directly affects health. According to Peltier, there has been a void in the science in this field (Courtesy UMass Amherst News Office, 2014). Development of new technologies that allow real-time evaluation of air pollution with toxicology variables rather than simply concentration monitoring of air pollutants is needed.

6.4 RISK ASSESSMENT AND MANAGEMENT

6.4.1 Air Emissions

The fine dust concentration in the Seoul metropolitan area in the early 2000s was very high, approximately 2–3.9 times higher than the other major cities in developed countries, and 1.3–1.4 times higher in other regions (noncapital regions) as well.

In order to protect the health of Korean citizens from such serious air pollution and to improve national development by reducing the social costs produced by air pollution, the government established and promoted a basic plan for air quality management in the capital area. The goal of this plan is to improve the fine dust pollution to the level of other developed countries. In this case, the sea in front of Incheon City can be seen from the top of Mt. Nam on a clear day. In the meantime, the fine dust concentration is improving every year by the continuous promotion of government policies such as installation of emission reduction devices on diesel engine vehicles and low-pollution vehicle supply projects (393,837 vehicles), total air pollution load management systems at places of business (304 places), and low NOx burner installation projects (2931 vehicles). The fine dust concentration in the capital area was reduced from 56 μg/m^3 in 2008 to 51 μg/m^3 in 2011, approximately a 9% reduction, and the number of days for which environmental requirements was satisfied increased from 175 days in 2008 to 205 days in 2011, approximately a 37% improvement. The average fine dust concentration of Seoul in 2011 was 44 μg/m^3, the lowest value since the time fine dust concentration has been measured.

The fine dust concentration of Seoul in 2010 was 47 μg/m^3, which satisfied the 50 μg/m^3 air quality standard for 2 consecutive years. The concentrations for Incheon and Gyeonggi were 51 and 53 μg/m^3, respectively, which were also the lowest-ever recorded values. Visibility is improving in all areas of Seoul, Incheon, and Gyeonggi Province.

VOCs are emitted from various sources, and organic solvents, the biggest contributor, account for 69.7% of all emissions; next are places of business such as manufacturers, which account for 14.3%. VOCs constitute approximately 25% of the air pollutants (ECOREA, 2013) (Figure 6.4).

VOC emission from solvent utilization for coatings increased annually until 2004, similar to the total VOC emission trend, but emissions were reduced in 2005 and then increased in 2006, by 30,000 tons (ECOREA, 2013) (Figure 6.5). Furthermore,

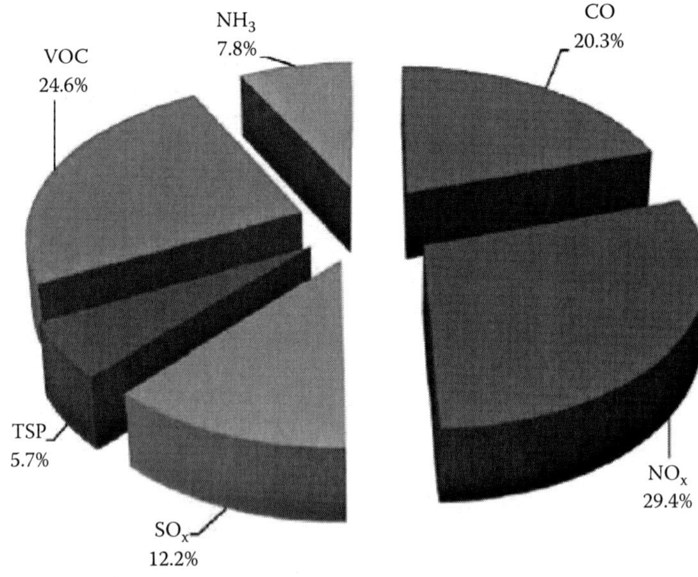

FIGURE 6.4 Contribution of the emissions of air pollutants in 2011.

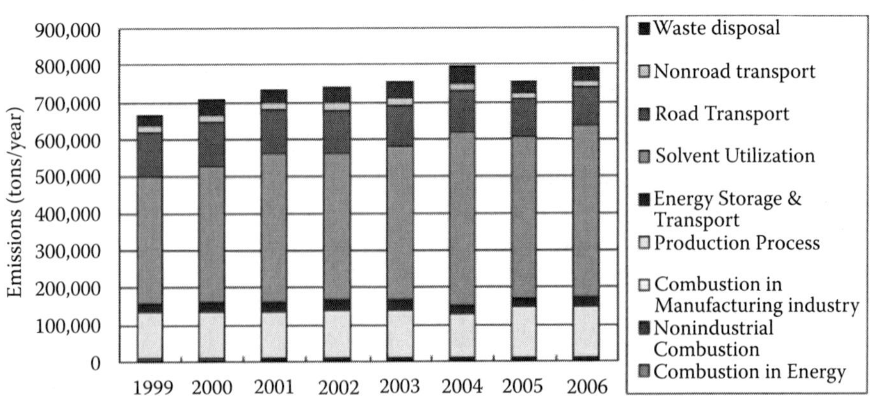

FIGURE 6.5 VOC emission trends by source categories from 1999 to 2006.

production processes and waste disposal contributed to the increases in VOC emission. Total VOC emissions in 2011 were 873,108 tons.

6.4.2 Air Policy Overview

Allowable emission standards for air pollutants emitted from discharging facilities regulate gaseous substances and particles separately starting from January 1, 2015, based on the definition of Separate Table 8 of the Enforcement Regulations of the Clean Air Conservation Act. This standard from the year 2015 will be separately applied by using a common standard or an enforced standard in case emission amounts of a single specified harmful substance exceeds tons per year.

For metropolitan cities, provinces, or special self-governing provinces, if it is recognized that the province environmental standard cannot be maintained or the air quality of an environmentally regulated area needs to be improved, more stringent allowable emission standards than the allowable emission standard in the Clean Air Conservation Act can be set.

If there are any areas where the allowable emission standard according to ordinances is not applied in the cities or provinces where the standard is applied, then the allowable emission standard according to ordinances will be applied to the discharging facility installed or to be installed in those areas.

The purpose of the total air pollution load management system is to improve the atmospheric environment of the capital area and the basis of this system is the Special Act on Seoul Metropolitan Air Quality Improvement, which was enacted in December 2003. The demonstration project was conducted in 100 business places from May 2006 to March 2007. After the total air pollution load management system was conducted in 118 Class 1 business places in July 2007, the management system was expanded to approximately 300 Class 2 business places in 2010. The target air pollutants for the management are NOx and SOx.

The total air pollution load management system is an advanced environment management system that controls the emissions of pollutants by converting the existing post concentration (ppm, mg/m^3) management to preventive total amount (kg) management, and the purpose of this system is to control the air pollution materials within the proper environment limit. This is an excellent environment management policy that reduces energy use and air pollutants simultaneously by inducing the pollutants to be emitted within the allocated amount by allocating the allowable emissions by year. This system is being executed separately for the local total maximum load of the metropolitan council and total maximum load of business places.

The total air pollution load management system for business places was planned to be conducted by setting standards for target materials such as NOx, SOx, and dust. However, for dust, because it is difficult to set an allocation standard of an emission amount and the contribution rate is very low compared with the whole emission amount, the allocation of total allowable emission amount was postponed in December 2007. Moreover, the capital area air environmental management committee decided that dust is exempted from the management target substances in August 2009. Dust was finally exempted from the target substances through a modification of the Special Law, and the system is being applied to NOx and SOx now. In addition, the enforcement plan for the allocation of total allowable emissions for business places was prepared during July through December of 2007 and was enforced on January 1, 2008, with 118 business places where the annual emission is more than 30 tons (NOx) and 20 tons (SOx).

6.4.3 Protecting Public Health from Environmental Risks

Since 2000, environmental attitudes have changed, focusing on the health effects of environmental pollution rather than controlling pollution. The Ministry of Environment established the 5-Year Comprehensive Plan for Environmental Health (2006–2015), which suggested a mid- and long-term planning blueprint to transit the policy protecting public health. In addition, the Environment Health Act was enacted in 2008. The purpose of the Environment Health Act is to investigate and audit damages and impacts of environmental pollution on public health and the ecosystem, to protect people from public health threats, and to maintain the health of people and the ecosystem. The ministry established the National Master Plan suggesting fundamental principles of an environmental and health policy and its detailed action plans that reflect circumstances and changes in policy and social needs. The Comprehensive Plan for Environmental Health (2011–2020) was established as a legal comprehensive plan according to the Environment Health Act.

The objectives of the Comprehensive Plan for Environmental Health (2011–2020) are to build a healthy and safe society by protecting the people's health from environmental risks and to leap into being a more advanced country in environmental health by reducing the burdens of environmental diseases. The Ministry of Environment is promoting 65 detailed projects in five areas, including the following projects: (1) minimization of environmental health risks, (2) reduction of health effects caused by different hazards, (3) establishment of a management system for controlling whole processes of environmental hazards, and (4) health protection from environmental risks. The 65 detailed projects in five areas promote investigation, surveillance, and damage aid of environmental diseases and preparation of groundwork for climate change for the next 10 years.

Urban air pollution causes morbidity- and mortality-related health damage and negatively affects environmental amenity benefits due to reduced visibility. Measurements of these losses from air pollution are important in order to justify environmental policy changes because the environmental impact involves a considerable cost to the government and because economic activities contribute to and are associated with air pollution (US Environmental Protection Agency [USEPA], 2000).

Various methods of environmental services valuation may be employed to estimate benefits from the reduction of urban air pollution. These include physical linkage methods involving estimation of dose–response functions in epidemiological studies, observed behavioral methods consisting of household health production function models, hedonic property price or wage models, and hypothetical behavioral or contingent valuation methods.

A large number of empirical studies are available that apply observed behavioral methods of valuation. This paper will attempt to evaluate the health damage cost due to exposure risks from particulate matter combined with risk assessment and contingent valuation methods for economic evaluation.

To perform this analysis, it is important to gather data about the environmental benefits that may be derived from estimating the value of risk reduction. Economists calculate the value of a reduction in risk of death by determining how much an individual is willing to pay for it. An individual's willingness to pay (WTP) to reduce a specific mortality risk is conventionally estimated by multiplying the change of probability of death by an estimate of the individual's marginal rate of substitution between wealth and mortality risk (Eeckhoudt and Hammitt, 2001). There are many methods for estimating WTP to reduce the mortality risk. One approach for measuring WTP is to infer the value from compensating wage differentials in the labor market. A second approach for measuring WTP is to ask people what they would pay to reduce their risk of dying; this is referred to as the contingent valuation method (Hanemann, 1985). Recently, the contingent valuation method has been increasingly used to estimate WTP.

In our study, we used the third approach to estimate a WTP amount for reducing a mortality risk. The third contingent valuation approach is often preferred, as it generates a scenario similar to that encountered by consumers in their usual market transactions (Lee et al., 2011).

This study evaluated the prospective damage costs of $PM_{2.5}$ inhalation. We performed a health risk assessment based on an exposure–response function to estimate the annual population risk in the Seoul metropolitan area. We also estimated a WTP amount for reducing the mortality rate in order to evaluate a statistical life value.

We combined the annual population risk and the value of statistical life to calculate the damage cost estimate. In the health risk assessment, we applied the $PM_{2.5}$ RR to evaluate the annual population risk. We targeted an exposure population of 5,401,369 persons who were over the age of 30 years.

Using a Monte Carlo simulation for uncertainty analysis, we estimated that the population risk of $PM_{2.5}$ inhalation during a year in Seoul is 2181 premature deaths for acute exposure and 18,510 premature deaths for chronic exposure. The monthly average WTP for 5/1000 mortality reduction over 10 years is USD 20.20 (95% CI, USD 16.60–24.50), and the implied value of statistical life is USD 485,000 (95% CI, USD 398,000–588,000). The damage cost estimate due to risk from $PM_{2.5}$ inhalation in Seoul is approximately USD 1057 million per year for acute exposure and USD 8972 million per year for chronic exposure.

It is important to note that this cost estimate does not reflect all health damage cost estimates in this urban area. This recommendation is a model for evaluating a mortality risk reduction, and as such, we must reevaluate an integrated application of morbidity risk.

In our results, the annual attributable number of deaths due to PM inhalation in Seoul is 2181 for acute exposure and 18,510 premature deaths for chronic exposure. The annual average number of deaths (including accidents and suicides) in Seoul is approximately 40,000, and the contribution of air pollution is quite large. In addition, the damage cost of PM inhalation is approximately USD 1000 million per year for acute and USD 9000 million for chronic exposure.

It is important to note that this cost estimate does not reflect all health damage cost estimates in this urban area. This recommendation is a model for evaluating a mortality risk reduction, and as such, we must reevaluate an integrated application of morbidity risk.

6.5 CONCLUSION

The ROK is one of the most industrialized countries in the world, and rapid industrialization has caused dramatic economic growth as well as air pollution. Numerous epidemiologic studies found that polluted air has adverse effects on Koreans' health, mainly on the cardiovascular and respiratory systems. However, epidemiologic studies in the ROK so far have concentrated on acute effects due to air pollution, and therefore, case-control or cohort studies are needed to investigate chronic effects. Furthermore, advanced methodology in exposure measurement would be helpful to lessen misclassification bias. Validity in epidemiologic studies and exposure measurement will support risk assessment and policy making and ultimately contribute to prevention of diseases related to environmental pollution.

REFERENCES

AirKorea. 2012. *We Broadcast Live the Air of Korea.* [Online]. Available: http://www.airkorea.or.kr/. Korea: Ministry of Environment.

AirKorea. 2014. *We Broadcast Live the Air of Korea.* [Online]. Available: http://www.airkorea.or.kr/. Korea: Ministry of Environment.

Asia Pacific Energy Centre. 2004. *Energy in China: Transportation, Electric Power and Fuel Markets*. Japan: Asia Pacific Energy Research Centre.

Braga, A. L., Saldiva, P. H., Pereira, L. A., Menezes, J. J., Conceicao, G. M., Lin, C. A., Zanobetti, A., Schwartz, J. and Dockery, D. W. 2001. Health effects of air pollution exposure on children and adolescents in Sao Paulo, Brazil. *Pediatric Pulmonology,* 31, 106–13.

Brook, R. D., Rajagopalan, S., Pope, C. A., III, Brook, J. R., Bhatnagar, A., Diez-Roux, A. V., Holguin, F., Hong, Y., Luepker, R. V., Mittleman, M. A., Peters, A., Siscovick, D., Smith, S. C., Jr., Whitsel, L. and Kaufman, J. D. 2010. Particulate matter air pollution and cardiovascular disease: An update to the scientific statement from the American Heart Association. *Circulation,* 121, 2331–78.

Chung, K. C. 1969. Epidemiological study of air pollution and its effects on health of urban population. *Korean Journal of Preventive Medicine,* 2, 5–22.

ECOREA. 2013. *Environmental Review 2013.* Korea: Korea Ministry of Environment.

Eeckhoudt, L. R. and Hammitt, J. K. 2001. Background risks and the value of a statistical life. *Journal of Risk and Uncertainty,* 23, 261–79.

Energy Korea. 2012. *Density of PM2.5 in Korea Shown to Be Twice as High as That of Major U.S. Cities*. Korea: Energy Korea.

Ha, E. H., Hong, Y. C., Lee, B. E., Woo, B. H., Schwartz, J. and Christiani, D. C. 2001. Is air pollution a risk factor for low birth weight in Seoul? *Epidemiology,* 12, 643–8.

Ha, E. H., Lee, J. T., Kim, H., Hong, Y. C., Lee, B. E., Park, H. S. and Christiani, D. C. 2003. Infant susceptibility of mortality to air pollution in Seoul, South Korea. *Pediatrics,* 111, 284–90.

Ha, K. H., Suh, M., Kang, D. R., Kim, H. C., Shin, D. C. and Kim, C. 2011. Ambient particulate matter and the risk of deaths from cardiovascular and cerebrovascular disease. *The Korean Society of Hypertension,* 17, 74–83.

Hanemann, W. M. 1985. Some issues in continuous—and discrete—response contingent valuation studies. *Northeastern Journal of Agricultural and Resource Economics,* 14, 5–13.

Hong, Y. C., Lee, J. T., Kim, H. and Kwon, H. J. 2002. Air pollution: A new risk factor in ischemic stroke mortality. *Stroke,* 33, 2165–9.

Hwang, S. S., Kang, S., Lee, J. Y., Lee, J. S., Kim, H. J., Han, S. K. and Yim, J. J. 2014. Impact of outdoor air pollution on the incidence of tuberculosis in the Seoul metropolitan area, South Korea. *Korean Journal of Internal Medicine,* 29, 183–90.

Kim, D.-S. 2013. Air pollution history, regulatory changes, and remedial measures of the current regulatory regimes in Korea. *Journal of Korean Society for Atmospheric Environment,* 29, 353–68.

Kim, J. B., Kim, C., Choi, E., Park, S., Park, H., Pak, H. N., Lee, M. H., Shin, D. C., Hwang, K. C. and Joung, B. 2012. Particulate air pollution induces arrhythmia via oxidative stress and calcium calmodulin kinase II activation. *Toxicology and Applied Pharmacology,* 259, 66–73.

Kim, J. H., Lim, D. H., Kim, J. K., Jeong, S. J. and Son, B. K. 2005. Effects of particulate matter (PM10) on the pulmonary function of middle-school children. *Journal of Korean Medical Science,* 20, 42–5.

Korea Environment Corporation. 2012. *Annual Air Quality Trends*. Korea: Korea Ministry of Environment.

Korea National Institute of Environmental Research. 2013. *Annual Report of Air Quality in Korea 2012*. Korea: Korea Ministry of Environment.

Korea National Statistical Office. 2008. *The Socioeconomic Changing Pattern Statistics in Republic of Korea during 60 Years*. Korea: Korea National Statistical Office.

Kwon, H. J. and Cho, S. H. 1999. Air pollution and daily mortality in Seoul. *Korean Journal of Preventive Medicine,* 32, 191–99.

Kwon, H. J., Cho, S. H., Nyberg, F. and Pershagen, G. 2001. Effects of ambient air pollution on daily mortality in a cohort of patients with congestive heart failure. *Epidemiology,* 12, 413–9.

Lee, J. T., Kim, H., Cho, Y. S., Hong, Y. C., Ha, E. H. and Park, H. 2003. Air pollution and hospital admissions for ischemic heart diseases among individuals 64+ years of age residing in Seoul, Korea. *Archives of Environmental Health,* 58, 617–23.

Lee, J. T., Kim, H., Song, H., Hong, Y. C., Cho, Y. S., Shin, S. Y., Hyun, Y. J. and Kim, Y. S. 2002. Air pollution and asthma among children in Seoul, Korea. *Epidemiology,* 13, 481–4.

Lee, J. T., Lee, S. I., Shin, D. and Chung, Y. 1998. Air particulate matters and daily mortality in Ulsan, Korea. *Korean Journal of Preventive Medicine,* 31, 82–90.

Lee, Y. J., Lim, Y. W., Yang, J. Y., Kim, C. S., Shin, Y. C. and Shin, D. C. 2011. Evaluating the PM damage cost due to urban air pollution and vehicle emissions in Seoul, Korea. *Journal of Environmental Management,* 92, 603–9.

Longo, L. D. 1977. The biological effects of carbon monoxide on the pregnant woman, fetus, and newborn infant. *American Journal of Obstetrics and Gynecology,* 129, 69–103.

Min, K. B., Min, J. Y., Cho, S. I. and Paek, D. 2008. The relationship between air pollutants and heart-rate variability among community residents in Korea. *Inhalation Toxicology,* 20, 435–44.

National Institute of Environmental Research. 2009. *Intercomparison Study from International Cooperative Research on Long-Range Transboundary Air Pollutants in Northeast Asia.* Seoul, Korea.

National Institute of Environmental Research. 2012. *Evaluation of the Long-Range Transboundary Air Pollutants in Northeast Asia (LTP) Project*. Seoul.

Park, M., Luo, S., Kwon, J., Stock, T. H., Delclos, G., Kim, H. and Yun-Chul, H. 2013. Effects of air pollution on asthma hospitalization rates in different age groups in metropolitan cities of Korea. *Air Quality, Atmosphere, and Health*, 6, 543–51.

Peters, A., Doring, A., Wichmann, H. E. and Koenig, W. 1997. Increased plasma viscosity during an air pollution episode: A link to mortality? *Lancet,* 349, 1582–7.

Sharma, A. P., Kim, K.-H., Ahn, J.-W., Sohn, J.-R., Lee, J.-H., Ma, C.-J. and Brown, R. J. C. 2014. Ambient particulate matter (PM10) concentrations in major urban areas of Korea during 1996–2010. *Atmospheric Pollution Research,* 5, 161–69.

Shin, D. C. 2010. *Evaluation of the Association between Air Quality Improvement and Diseases*. Seoul: Citizen's Committee for Clean Seoul.

Son, J. Y., Bell, M. L. and Lee, J. T. 2010. Individual exposure to air pollution and lung function in Korea: Spatial analysis using multiple exposure approaches. *Environmental Research,* 110, 739–49.

Son, J. Y., Bell, M. L. and Lee, J. T. 2011. Survival analysis of long-term exposure to different sizes of airborne particulate matter and risk of infant mortality using a birth cohort in Seoul, Korea. *Environmental Health Perspective,* 119, 725–30.

United Nations. 2012. *World Urbanization Prospects: The 2011 Revision,* New York, Population Division, Department of Economic and Social Affairs, United Nations.

UMass Amherst News Office. 2014. *Peltier Forges New Methods for Measuring Air Pollution* [Online]. Available: https://www.umass.edu/sphhs/news-events/peltier-forges-new-methods-measuring-air-pollution.

US Environmental Protection Agency. 2000. *Guideline for Preparing Economic Analysis.* Washington DC: National Center for Environmental Economics, Office of Policy, US Environmental Protection Agency.

Yi, O., Hong, Y. C. and Kim, H. 2010. Seasonal effect of PM (10) concentrations on mortality and morbidity in Seoul, Korea: A temperature-matched case-crossover analysis. *Environmental Research,* 110, 89–95.

Section II

Beyond What We Have Known

7 Air Pollution and Mental Health

Changsoo Kim and Jaelim Cho

CONTENTS

7.1 INTRODUCTION

It has been widely known that air pollution affects pulmonary and cardiovascular health and has been the focus of research. Of late, novel epidemiological findings of air pollution effects on various organs have been reported. The brain is noteworthy among the organs, since it might not be easy for the general public to intuitively accept the relationship between air pollution and brain health. However, it appears that biological evidence is sufficient to warn about adverse effects of air pollution exposure on the brain. Particulate matter can influence the central nervous system (CNS) in two major ways: causing neuroinflammation directly through the olfactory nerve (ultrafine particles) and indirectly through systemic inflammation, which can be accompanied by cytokine production, microglia activation, lipid peroxidation, and neuron damage in the brain (Block and Calderon-Garciduenas, 2009). These neuroinflammatory mechanisms may result in changes in neurotransmitters such as serotonin and dopamine (Gonzalez-Pina and Paz, 1997) and the accumulation of amyloid beta (Aβ)-42 (a marker related to Alzheimer's disease) and alpha-synuclein (a marker related to Parkinson's disease) (Calderon-Garciduenas et al., 2008). According to an experimental study, particulate matter and ozone exposure can lead to an increase in stress hormone, which is linked to the activation of the hypothalamic–pituitary–adrenal (HPA) axis (Thomson et al., 2013). As cardiac arrhythmia is known to be related to air pollution, the effect of air pollution on the autonomic nervous system (ANS) is also of interest. In particular, ozone causes local inflammation in the lungs, followed by *nucleus tractus solitarius* neuron activation through the vagal nerves, which are related to the ANS (Gackiere et al., 2011). Supporting this biological evidence, numerous epidemiological studies have suggested associations of air pollution with (1) suicidal behaviors (attempted suicide and completed suicide), (2) psychiatric disorders (depression and anxiety), (3) neurodegenerative diseases (cognitive decline, dementia, and Parkinson's disease), and (4) neurodevelopmental disorders (attention deficit hyperactivity disorder [ADHD] and autism). In this chapter, major epidemiological findings of each category are summarized.

7.2 SUICIDAL BEHAVIORS

Suicide is one of the most complicated mental health issues. The World Health Organization (WHO) estimates that approximately one million people die by suicide each year (WHO, 2005). The term "suicidal behaviors" refers to suicide attempt and completed suicide (i.e., death from suicide), and it has been known that suicide attempt and completed suicide have divergent characteristics. Both are prevalent in young adults, but men are at higher risk for completed suicide, whereas women are more likely to attempt suicide. In general, suicide attempters use less lethal methods such as poisoning while suicide completers harness lethal methods such as hanging and firearms. Suicide is involved in a variety of factors encompassing mental disorders (depression, bipolar disorder, schizophrenia, and anxiety disorders), environmental circumstances (access to means for suicide, chronic stress, and stressful life events), and historical factors (family history of suicide and mental disorders). In addition to these individual factors, ecological factors such as unemployment rate and celebrity suicide have been reported in the literature, but environmental factors, especially air pollution, have been relatively marginalized. So far, only a small number of investigations on the relationship between environmental pollution and suicidal behaviors were conducted by researchers in several countries.

Szyszkowicz et al. (2010) reported on the relationship between carbon monoxide (CO), nitrogen dioxide (NO_2), and particulate with the diameter of 10 μm or less (PM_{10}) exposure and emergency department visits for suicide attempt in Vancouver, Canada. With time-series data, the authors harnessed the generalized linear mixed models on clusters, in which days of the same day of the week in the same month and year were bracketed. As the authors described, the number of emergency department visits for suicide attempt was low and suicide mortality might have been more appropriate.

In the Republic of Korea, deaths from suicide soared in early 2000s; thus, our research team took notice of the remarkable increase in suicide mortality in the middle-aged and elderly, indicating a rising prevalence of late-onset depression. In particular, the middle-aged and elderly with cardiovascular disease might be at risk of late-onset depression through the development of vascular depression. As many studies reported, cardiovascular disease such as myocardial infarction is associated with ambient air pollution through the mechanism of systemic inflammation. We therefore hypothesized that air pollution had a short-term effect on suicide, especially among those with underlying cardiovascular disease (Kim et al., 2010). We obtained particulate matter data from automatic monitoring systems of seven big cities in the Republic of Korea and data from suicide completers in 2004. To identify underlying diseases, data of suicide completers were merged with medical care utilization data based on their national health insurance claims. Using a time-stratified case-crossover study design, we found a 9.1% greater risk of suicide due to PM_{10} exposure (interquartile range 27.59 μg/m^3), and an 18.9% greater risk in patients with preexisting cardiovascular disease.

In Taiwan, Yang et al. (2011) suggested that the pattern of rising suicide mortality was predicted by increasing PM_{10}. Sulfur dioxide (SO_2) and ozone (O_3) also increased the risk for suicide with a longer secular trend. After conducting empirical mode decomposition for decomposing time-series data of suicide mortality and for detrending air pollution, weather, and unemployment data, the authors performed multiple linear regression analysis. Interestingly, this study made a distinction between violent suicides (e.g., hanging and jumping from heights) and nonviolent suicides (e.g., poisoning) and showed that air pollution contributed more to nonviolent suicides than violent suicides. As suggested by this study, the incidence of suicide is known to be affected by meteorological variables. However, meteorological factors vary across geographical regions, where various characteristics of population, including demographic distribution and cultural aspects, may lead to inconsistent results. Comparing the characteristics of suicide between Asian and Western countries suggests that Asians are more likely to commit impulsive suicidal behaviors (WHO, 2009). Impulsiveness of suicidal behaviors may affect the choice of suicide methods, which more likely consist of easier-access means. Because of these complex aspects of suicide, it might be inevitable to see a discrepancy between results from different countries. In the United States, Bakian et al. (2015) reported associations between exposure to NO_2 and $PM_{2.5}$ and suicide mortality in Salt Lake County, Utah, in 2000–2010. Dissimilar to the Taiwanese study, this study presented the significant association only for violent suicides. The study by Bakian et al. categorized drug overdose, drowning, poisoning, and gas into nonviolent suicides, while the study by Yang et al. included jumping from heights in the violent suicides. It is possible that jumping from heights includes drowning, so the different categorization might have influenced this discrepancy. However, it is difficult to exclude the possibility of racial and cultural differences. Thus, a result from a city or a country may be difficult to be generalized to other populations; therefore, there is a need of more replication studies in various regions.

The link between air pollution and suicide behaviors suggested by above studies does not secure a causal relationship. Moreover, these studies focused on the short-term effect of air pollution. Therefore, it would be more adequate to describe that polluted air "triggers" suicidal behaviors in individuals with predisposing factors for suicidal behaviors. Despite the criticism that the finding is just a statistical inference when considering the complexity of suicide mechanism, these air pollution studies could underscore the adverse effects of ambient particulate matter on the brain in that they might be able to induce suicidal behaviors.

7.3 PSYCHIATRIC DISORDERS

7.3.1 Depression

It is estimated that almost 350 million people are suffering from depression globally, according to the WHO statement (WHO, 2012). Unipolar depressive disorder, which was the third leading cause of disease burden in 2004, is expected to become the primary cause of disease burden worldwide by 2030 (WHO, 2008). Depression or depressive disorder is characterized by depressed mood, decreased interest, fatigue, guilt/worthlessness, suicidal ideation, decreased ability to concentrate, and changes in weight, sleep, and activity. Depression can present in various ways: major depression, psychotic depression (occurring with psychotic symptoms), seasonal depression, and postpartum depression. According to the *Diagnostic and Statistical Manual of Mental Disorders*, 4th Edition (DSM-IV), when one of the first two symptoms are persistent for more than 2 weeks and at least five of the nine symptoms above are present nearly every day, a patient is diagnosed with major depressive disorder; additionally, patients likely have social, occupational, and educational impairments. The biological mechanism of depression has known to be mainly related to a decrease in neurotransmitters (e.g., serotonin, norepinephrine, dopamine, and gamma-aminobutyric acid). The dysregulation of the HPA axis is also involved in the development of depression because it may lead to sustained high levels of stress hormones like the corticotropin-releasing hormone, which is toxic to various brain structures such as the hippocampus.

Szyszkowicz et al. (2009) analyzed 27,047 emergency department visits for depression in six cities in Canada. The generalized linear mixed model with Poisson distribution was used, with temperature and relative humidity as covariates. CO and NO_2 were associated with increased daily emergency department visits for depression in the warm period (April–September). For PM_{10}, the largest increase, 7.2% (95% confidence interval [CI]: 3.0%–11.6%) per 19.4 $\mu g/m^3$, was observed in the cold period (October–March).

In the Republic of Korea, Lim et al. (2012) assessed 537 elders for depressive symptoms using the Korean version of the Geriatric Depression Scale-Short Form (SGDS-K). In the results from generalized estimating equations (GEE) analyses, SGDS-K scores were positively associated with interquartile range increases in the 3-day moving average concentration of PM_{10} (17.0%; 95% CI: 4.9%–30.5%), the 0–7-day moving average of NO_2 (32.8%; 95% CI: 12.6%–56.6%), and the 3-day moving average of O_3 (43.7%; 95% CI: 11.5%–85.2%).

Similar to the study by Lim et al. in the Republic of Korea, Wang et al. (2014) assessed depressive symptoms among participants in the Maintenance of Balance, Independent Living, Intellect, and Zest in the Elderly of Boston (MOBILIZE Boston) study using another depression screening tool, the Revised Center for Epidemiological Studies Depression Scale (CESD-R). They used residential proximity to the nearest major road as a proxy of long-term exposure to traffic-related pollution and assessed short-term exposure to $PM_{2.5}$, sulfates, black carbon, ultrafine particles, O_3, CO, NO, and NO_2 (2-week averages before evaluation). GEE was used to estimate the odds ratio of a CESD-R score ≥ 16 associated with exposure. There was no evidence of a positive association between depressive symptoms and long-term exposure to traffic pollution or short-term exposure to ambient air pollution. The authors discussed that the discrepancy with the Korean study might have been derived from the different depression screening tool and less severe air pollution in Boston. In addition, the prevalence of depressive symptoms in their samples was less than 10%, which limited the statistical power of the study.

Following our suicide research, we targeted mental disorders, which are one of the risk factors for suicide. Among mental disorders, depression is regarded as the major mental disorder leading to suicidal behavior. Moreover, considering the possible biological mechanism of "vascular depression" suggested by our previous study on suicide, it seemed plausible that air pollution could induce depression. Thus, we conducted a time-stratified case-crossover study revealing the association between air pollution and depression (Cho et al., 2014).

Although a previous Canadian study used emergency department data, one major question was how to define "depression" and what kind of data could be used in our setting when designing our study. Because the Republic of Korea adopted the system of national health insurance, researchers can obtain the information on medical care utilization from national health insurance claims data through the Health Insurance Review and Assessment (HIRA). The HIRA data contain the date of visit, the code of diagnosis based on the International Classification of Diseases 10th revision (ICD-10), clinic or hospital code, the duration of care, and the type of visit (outpatient, emergency department visit, and inpatient). In this study, we aimed to investigate the short-term effect of air pollution on the occurrence of depressive symptoms. In this context, outpatient and inpatient data were deemed to be inappropriate because the data included scheduled visits, in which we could not ascertain whether a depressive symptom occurred on the date of a patient's visit. Therefore, the most reasonable option was emergency department visit data, which can moderately reflect time and place when depressive symptoms started. Although there was a concern that depressive patients might rarely visit emergency departments, nearly 5000 cases of depressive episode (F32) were observed during the 5-year study period in Seoul.

In the results of the conditional logistic regression, SO_2, PM_{10}, NO_2, and CO were positively associated with emergency department visits for depressive episode: the maximum risk was observed in the distributed lag 0–3 model for PM_{10} (odds ratio, 1.120; 95% CI: 1.067–1.176). Interestingly, O_3 did not show any significance, which might result from the fact that O_3 is a secondary air pollutant generated from primary air pollutants such as NO_2 via photochemical reaction. Regarding this notion, there is a possibility that the primary air pollutants might trigger depressive symptoms before O_3 level starts to rise.

This single city study had several limitations, one of which was that we used data only from Seoul. Seoul is the capital of the Republic of Korea and has higher air pollution levels compared with other cities. This raises the question if the association between air pollution and depression might exist only in a city with severe air pollution. To look into this possibility of threshold, we illustrated curve graphs showing the risk estimates according to air pollution levels. It seemed that there was no significant threshold, suggesting that the effect of air pollution on depression might be found even in a city with a low level of air pollution.

In our previous study on suicide, we suggested a higher risk in individuals with preexisting cardiovascular illness. In this study, in addition to cardiovascular disease, the most interesting finding was a significantly higher risk for emergency department visits for depressive episode in individuals with underlying diabetes mellitus or asthma. Regarding this finding, we hypothesized that air pollution has an interactive effect on neuro-psycho-endocrine-immune connections via an inflammatory process: air pollution might function as an effect modifier in the bidirectional associations of depression with diabetes (endocrine disease) or allergic asthma (immune disease). This speculation is consistent with the finding of chronic obstructive pulmonary disease, without any significant results. However, we could not obtain information on prescribed medication such as steroids in the setting of this study, so further research is required. Nonetheless, the findings of this research indicated the population potentially susceptible to air pollution in terms of their mental health, which may have public health implications.

7.3.2 Anxiety

As mentioned earlier, the biological mechanism of air pollution effects on suicide and depression involves neuroinflammation via systemic inflammation. There have been numerous animal experiments showing the process that can affect brain regions, such as the amygdala, and that can result in changes in levels of neurotransmitters such as dopamine and serotonin (Gonzalez-Pina and Paz, 1997). Together with depression, anxiety is related to these neurotransmitters and brain regions that may be affected by air pollution, and depression and anxiety share a common pathophysiology. Nonetheless, the effect of ambient air pollution on anxiety has been little investigated. Thus, we performed a time-series study to identify the association (Cho et al., 2015) but encountered the first question about how to define anxiety. Anxiety disorders include phobic anxiety, in which the symptom trigger factor is situational and specific; thus, if we encompassed all anxiety symptoms, we would observe nothing significant. Therefore, we focused on nonspecific and spontaneous anxiety, which is a major feature of panic disorder, and used emergency department visit data. Although there were only 2320 cases of panic disorder in emergency department visits in Seoul over 5 years, a slight but significant increased risk was observed in our time-series study. Interestingly, the significant association was found only for O_3. Previous occupational studies have reported the occurrences of panic symptom in workers dealing with organic solvents, and these authors suggested that the major route of exposure was inhalation. According to the Haagen-Smit theory, O_3 can be generated secondly by photochemical reaction with primary pollutants such as volatile organic compounds (VOCs). Collectively, the association of panic disorder with O_3 exposure suggests that further measurement of VOCs would be helpful to examine this speculation. Nonetheless, the association between O_3 exposure and panic disorder seems biologically plausible according to an animal study suggesting that O_3 induces strong local inflammation in the lungs, which can reach the brain via the vagus nerve, indicating the potential effect of O_3 on the ANS. This consecutive process includes the involvement of the *nucleus tractus solitarius*, which is a central chemoreceptor and respiratory center, in the *medulla oblongata*. If this region is affected, the respiratory center can be erroneously activated, causing respiratory symptoms such as hyperventilation and shortness of breath. According to the false suffocation alarm theory, this mechanism plays a role in the development of panic disorder with respiratory symptoms. Panic disorder with respiratory symptoms, as one of the subtypes of panic disorder, is more likely to induce spontaneous attacks and is distinctive from panic disorder without respiratory symptoms in terms of characteristics. Our study targeted spontaneous development of panic symptoms; thus, the possible mechanism we suggested was consistent with our finding that the increased risks remained significant only in women, who are more likely to suffer from the respiratory subtype of panic disorder.

The most questionable aspect is the interactive effect of temperature on the relationship between O_3 and panic disorder. After stratification of season, we found that the risks were higher in spring and summer and lower in autumn and winter. This might be related to higher average temperature and ambient O_3 concentrations in spring and summer compared with the other seasons. Thus, we assumed that there could be a threshold of O_3 concentration that can trigger panic disorder and found a rough value through further analysis. There might be a synergistic effect of temperature and O_3 exposure, or temperature itself might be associated with panic disorder, so further studies would be required. Although many epidemiological studies on the effect of air pollution on the brain have suggested the biological mechanism of neuroinflammation in the CNS, this study highlighted the O_3 effect on the ANS, leading to the development of panic symptoms. Additional nervous diseases can be related to the effect of air pollution on the ANS, which deserve further investigations.

While our study focused on the short-term effect of ambient O_3 exposure on nonphobic anxiety, Power et al. (2015) suggested a long-term effect of $PM_{2.5}$ exposure on high symptoms of phobic anxiety. Among 71,271 women registered in the Nurses' Health Study, subjects with the phobic anxiety subscale of the Crown-Crisp index score ≥ 6 were considered as having high symptoms of anxiety (15%). Having high anxiety symptoms was associated with higher exposure to $PM_{2.5}$; odds ratios were 1.12 (95% CI: 1.06–1.19) and 1.15 (95% CI: 1.06–1.26) for the previous 1-month and 12-month exposure periods, respectively.

Despite a growing body of literature, how air pollutants contribute to the development of depression and anxiety symptoms is poorly understood. Considering results from several experimental and epidemiological studies, the possible mechanisms mentioned earlier are summarized in Figure 7.1.

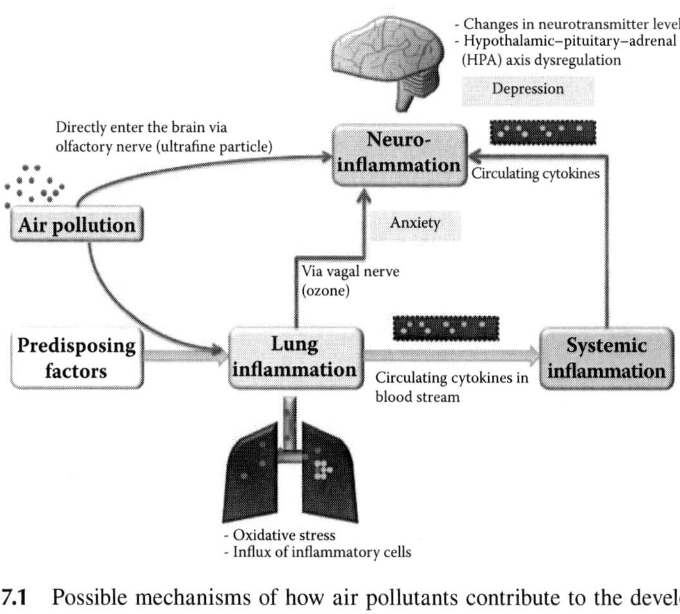

FIGURE 7.1 Possible mechanisms of how air pollutants contribute to the development of depression and anxiety symptoms.

7.4 NEURODEGENERATIVE DISEASES

Neurodegeneration refers to progressive neuronal loss, including functional and structural loss. This process results in the development of neurodegenerative diseases such as dementia, Parkinson's disease, and amyotrophic lateral sclerosis (ALS).

Dementia is the umbrella term for chronic and progressive loss of memory and thinking ability. The number of dementia patients globally is currently estimated at 47.5 million and is expected to increase to 75.6 million by 2030, and more than triple by 2050 (WHO and Alzheimer's Disease International, 2012). The major causes of dementia are Alzheimer's disease, vascular dementia, and frontotemporal dementia. Alzheimer's disease accounts for the largest body of dementia. Known risk factors for Alzheimer's disease include age, heredity, family history, and previous head injuries. What causes Alzheimer's disease is not fully understood, but the pathophysiology involves the ε4 allele of the apolipoprotein E (ApoE), Aβ deposits, and tau protein abnormalities forming neurofibrillary tangles. Cytokines and inflammatory responses may also play a role in the pathophysiology.

Parkinson's disease is a neurodegenerative disease of the CNS, in particular the movement system. The pathophysiology involves the death of dopamine-generating cells in the *substantia nigra*, a part of the midbrain. Symptoms include pill-rolling tremor, cogwheel rigidity, and slowness of movement at the early stage; and eventually dementia develops at the advanced stage. The motor symptoms are referred to as parkinsonism, which is classified as primary and secondary according to its cause. Primary parkinsonism has no obvious cause (i.e., idiopathic) but may have a genetic origin. Secondary parkinsonism results from preexisting health problems such as brain injury, stroke, and chemical or drug poisonings. As for the risk and protective factors of Parkinson's disease, exposure to pesticides is the best-known risk factor, but the role of heavy metals and the protective effect of tobacco smoking are in controversy.

Symptoms of ALS involve motor neurons and muscles and include stiffness, twitching, wasting, and gradually worsening weakness of muscles, followed by difficulty moving, speaking, swallowing, and finally breathing. It is known that about 10% of cases are related to heredity (Kiernan et al., 2011) while the remainder have an unclear etiology.

The short-term effect of air pollution on mental health deserves more attention in terms of environmental health, but it would be difficult to apply the results directly to environmental health policy because the associations are diminutive and still in controversy—especially when it comes to the causal relationship. Therefore, to investigate the causal relationship between air pollution and mental health, it is imperative to perform a longitudinal study on the long-term effects with a careful design. Moreover, risk assessment, which is useful in environmental health policy establishment, should be accompanied by the process of identifying dose–response relationship, which can be obtained by conducting a longitudinal study. With regard to mental health problems such as neurodegenerative diseases, the pathological process develops slowly, and the symptoms are more likely to be detectable when the diseases have progressed, which makes it less plausible to examine the short-term effect of air pollution on the occurrence of neurodegenerative diseases. So far, several studies linked long-term exposure to air pollution with the incidence of neurodegenerative diseases. Finkelstein and Jerrett reported the relation between ambient manganese levels and Parkinson's disease (Finkelstein and Jerrett, 2007). The authors established a cohort of 110,348 subjects in Hamilton and Toronto and ascertained Parkinson's cases by linking administrative records. In Hamilton, the odds ratio for Parkinson's disease was 1.034 (95% CI: 1.00–1.07) per 10 ng/m^3 increase in manganese in total suspended particulate, whereas no association was found in Toronto. For ALS, Malek et al. (2015) suggested an association with ambient aromatic solvents in 12 counties of the United States in 1999 and 2002 through conducting a case-control study with 51 cases and 51 matched controls. However, the other neurotoxic hazardous air pollutants (HAPs) did not show significances.

In conducting studies on neurodegenerative diseases, one of the major challenges is how to define the outcome. Neurodegenerative diseases such as dementia have a relatively low prevalence compared with other chronic diseases such as cardiovascular disease and diabetes mellitus despite their growing number for decades. In addition, the effect of environmental pollution is more likely to be slight, which makes it more difficult to detect an association. Therefore, obtaining a sufficiently large sample size is needed to obtain significance but could be demanding in terms of time and cost unless administrative data are used as Finkelstein and Jerrett did. Considering that, measuring brain function will be more useful rather than using the incidence of disease as an outcome. Moreover, for early intervention, it would be worthwhile to find subtle brain changes before symptoms are evident. In line with this, several epidemiological studies have reported the association of air pollution with cognitive function among cohort subjects.

Power et al. (2011) reported the association between black carbon and cognitive function among 680 older men (50–99 years) from the US Department of Veterans Affairs Normative Aging Study. Doubling in black carbon was associated with having a mini–mental state examination (MMSE) score ≤ 25 (odds ratio, 1.3; 95% CI: 1.1–1.6]. For global cognitive function, doubling in black carbon was inversely associated with the cognitive test score.

With 19,409 older US women (70–81 years) from the Nurses' Health Study Cognitive Cohort, Weuve et al. (2012) associated recent (1 month) and long-term (7–14 years) exposures to $PM_{2.5-10}$ and $PM_{2.5}$ with cognitive decline based on the telephone interview for cognitive status. The authors analyzed the exposure levels before the baseline cognitive test and three repeated measures of cognitive function with GEE and found that a decrease in global cognitive score per 2 years was associated with higher $PM_{2.5-10}$ and $PM_{2.5}$ exposure.

In the United Kingdom, Tonne et al. (2014) found that PM_{10} and $PM_{2.5}$ were associated with cognitive decline in 2867 participants (mean age, 66; standard deviation, 6 years) of the Whitehall II cohort who were residents of Greater London. The authors assessed four cognitive functions (reasoning, memory, semantic fluency, phonemic fluency) twice with a 5-year interval. For the second assessment, a 5-year exposure to higher $PM_{2.5}$ exposure was associated with lower cognitive scores on reasoning. When excluding movers out of Greater London between the two assessments, exposure 4 years prior to the second assessment was significantly associated with a decline in memory score.

Because Alzheimer's disease is related to the ε4 allele of the ApoE, there may be interactive effects of the genotype on the relationship between air pollution and cognitive function. Schikowski et al. (2015) investigated the effect modification of the ApoE gene variants among 789 elderly women (mean age, 73; standard deviation, 3 years) from the Study on the influence of Air pollution on Lung function, Inflammation, and Aging (SALIA) cohort in Germany. The researchers assessed subjects' cognitive function using the Consortium to Establish a Registry for Alzheimer's Disease (CERAD) neuropsychological assessment battery. In this study, traffic load was inversely associated with the score on the figure-copying subtest in women with one or more ApoE ε4 risk allele but not in women without this risk allele.

Previous studies mentioned above used questionnaire-based tools for the evaluation of cognitive function. Questionnaire-based tests can identify functional changes of the brain, which may be preceded by subtle anatomical changes (Jack et al., 2013). To detect fine anatomical changes in the brain, it would be helpful to perform brain magnetic resonance imaging (MRI) and analyze the images to identify any microbleeds or estimate cortical thickness. A recent study by Wilker et al. (2015) associated long-term air pollution exposure with brain MRI findings such as total cerebral brain volume, hippocampal volume, white matter hyperintensities, and covert brain infarct. In this study, long-term exposure to higher $PM_{2.5}$ was associated with smaller total cerebral brain volume and more covert brain infarcts in 943 older (aged ≥ 60 years) adults residing in the New England region. Since 2014, our group also has conducted a cohort study (Environmental Pollution Induced Neurological Effect, EPINEF study) targeting brain structural and functional changes in relation to exposure to various air pollutants, using brain MRI and neuropsychological assessment. Despite its beginning stage, the multidisciplinary approach encompassing environmental epidemiology and brain science technology will be useful for detecting diminutive brain changes associated with air pollution and for identifying the underlying biological mechanism.

7.5 NEURODEVELOPMENTAL DISORDERS

Neurodevelopmental disorders refer to a constellation of neuropathological conditions occurring during developmental periods, which are accompanied by personal, social, educational, or occupational impairments. Symptoms in neurodevelopmental disorders range from specific deficits of learning or social communication to global deteriorations of intelligence. There are two well-known neurodevelopmental disorders: ADHD and autism spectrum disorder. ADHD is characterized by inattentiveness, impulsivity, and overactivity. Children with autism spectrum disorder have deficits of social skills, excessive repetitive behaviors, limited interests, and craving sameness. These disorders have a strong genetic and familial tendency, but it is widely accepted that both genetics and environmental factors may play a role. Recently, literature has accumulated on environmental factors for ADHD and autism, outdoor air pollution in particular. Exposure to traffic-related air pollution (TRAP) and heavy metals during prenatal or early-life period is considered as a potential culprit for developing ADHD and autism.

Newman et al. (2013) investigated the relationship between early-life exposure to TRAP (elemental carbon attributed to traffic [ECAT] as a surrogate) and ADHD symptoms at 7 years of age among 576 children of the Cincinnati Childhood Allergy and Air Pollution Study (CCAAPS) birth cohort. They found that exposure to the highest tertile of ECAT during infancy was significantly associated with hyperactivity T-scores > 59, and the association was stronger in mothers with higher education.

Volk et al. (2013) conducted a case-control study including 279 children with autism and 245 controls. In this study, exposure to TRAP, NO_2, $PM_{2.5}$, and PM_{10} during pregnancy and infancy was associated with autism. Further, Raz et al. (2015) suggested that higher exposure to $PM_{2.5}$ during pregnancy, the third trimester in particular, was associated with an increased risk for autism spectrum disorder through conducting a nested case-control study of participants from the Nurses' Health Study II (NHS II).

Several epidemiological studies evaluated preclinical changes in psychomotor and cognitive development. A study by Guxens et al. (2014) found that NO_2 and $PM_{2.5}$ exposure during pregnancy was inversely associated with psychomotor development, but not with cognitive development, among 9482 children from six European birth cohorts. Prenatal polycyclic aromatic hydrocarbons (PAHs) exposure is also of interest, and a brain imaging study showed that PAHs were associated with reduced white matter surface in children aged 7–9 years (Peterson et al., 2015).

7.6 CONCLUSIONS

In this chapter, we summarized major epidemiological findings on the relationship between air pollution and mental health problems: suicidal behaviors, depression, anxiety, neurodegenerative diseases, and neurodevelopmental disorders. These findings are supported by numerous experimental studies suggesting the biological plausibility; thus, concern of the effect of air pollution on the brain is growing. Although the results from epidemiological studies are heterogeneous, the growing evidence suggests that research is heading toward a better understanding of mental disorders with poorly known causes.

REFERENCES

Bakian, A. V., Huber, R. S., Coon, H., Gray, D., Wilson, P., Mcmahon, W. M., and Renshaw, P. F. 2015. Acute air pollution exposure and risk of suicide completion. *Am J Epidemiol*, 181, 295–303.

Block, M. L. and Calderon-Garciduenas, L. 2009. Air pollution: Mechanisms of neuroinflammation and CNS disease. *Trends Neurosci*, 32, 506–16.

Calderon-Garciduenas, L., Solt, A. C., Henriquez-Roldan, C., Torres-Jardon, R., Nuse, B., Herritt, L., Villarreal-Calderon, R., Osnaya, N., Stone, I., Garcia, R., Brooks, D. M., Gonzalez-Maciel, A., Reynoso-Robles, R., Delgado-Chavez, R., and Reed, W. 2008. Long-term air pollution exposure is associated with neuroinflammation, an altered innate immune response, disruption of the blood-brain barrier, ultrafine particulate deposition, and accumulation of amyloid beta-42 and alpha-synuclein in children and young adults. *Toxicol Pathol*, 36, 289–310.

Cho, J., Choi, Y. J., Sohn, J., Suh, M., Cho, S. K., Ha, K. H., Kim, C., and Shin, D. C. 2015. Ambient ozone concentration and emergency department visits for panic attacks. *J Psychiatr Res*, 62, 130–5.

Cho, J., Choi, Y. J., Suh, M., Sohn, J., Kim, H., Cho, S. K., Ha, K. H., Kim, C., and Shin, D. C. 2014. Air pollution as a risk factor for depressive episode in patients with cardiovascular disease, diabetes mellitus, or asthma. *J Affect Disord*, 157, 45–51.

Finkelstein, M. M. and Jerrett, M. 2007. A study of the relationships between Parkinson's disease and markers of traffic-derived and environmental manganese air pollution in two Canadian cities. *Environ Res*, 104, 420–32.

Gackiere, F., Saliba, L., Baude, A., Bosler, O., and Strube, C. 2011. Ozone inhalation activates stress-responsive regions of the CNS. *J Neurochem*, 117, 961–72.

Gonzalez-Pina, R. and Paz, C. 1997. Brain monoamine changes in rats after short periods of ozone exposure. *Neurochem Res*, 22, 63–6.

Guxens, M., Garcia-Esteban, R., Giorgis-Allemand, L., Forns, J., Badaloni, C., Ballester, F., Beelen, R., Cesaroni, G., Chatzi, L., De Agostini, M., De Nazelle, A., Eeftens, M., Fernandez, M. F., Fernandez-Somoano, A., Forastiere, F., Gehring, U., Ghassabian, A., Heude, B., Jaddoe, V. W., Klumper, C., Kogevinas, M., Kramer, U., Larroque, B., Lertxundi, A., Lertxuni, N., Murcia, M., Navel, V., Nieuwenhuijsen, M., Porta, D., Ramos, R., Roumeliotaki, T., Slama, R., Sorensen, M., Stephanou, E. G., Sugiri, D., Tardon, A., Tiemeier, H., Tiesler, C. M., Verhulst, F. C., Vrijkotte, T., Wilhelm, M., Brunekreef, B., Pershagen, G., and Sunyer, J. 2014. Air pollution during pregnancy and childhood cognitive and psychomotor development: Six European birth cohorts. *Epidemiology*, 25, 636–47.

Jack, C. R., Jr., Knopman, D. S., Jagust, W. J., Petersen, R. C., Weiner, M. W., Aisen, P. S., Shaw, L. M., Vemuri, P., Wiste, H. J., Weigand, S. D., Lesnick, T. G., Pankratz, V. S., Donohue, M. C., and Trojanowski, J. Q. 2013. Tracking pathophysiological processes in Alzheimer's disease: An updated hypothetical model of dynamic biomarkers. *Lancet Neurol*, 12, 207–16.

Kiernan, M. C., Vucic, S., Cheah, B. C., Turner, M. R., Eisen, A., Hardiman, O., Burrell, J. R., and Zoing, M. C. 2011. Amyotrophic lateral sclerosis. *Lancet*, 377, 942–55.

Kim, C., Jung, S. H., Kang, D. R., Kim, H. C., Moon, K. T., Hur, N. W., Shin, D. C., and Suh, I. 2010. Ambient particulate matter as a risk factor for suicide. *Am J Psychiatry*, 167, 1100–7.

Lim, Y. H., Kim, H., Kim, J. H., Bae, S., Park, H. Y., and Hong, Y. C. 2012. Air pollution and symptoms of depression in elderly adults. *Environ Health Perspect*, 120, 1023–8.

Malek, A. M., Barchowsky, A., Bowser, R., Heiman-Patterson, T., Lacomis, D., Rana, S., Ada, Y., and Talbott, E. O. 2015. Exposure to hazardous air pollutants and the risk of amyotrophic lateral sclerosis. *Environ Pollut*, 197, 181–6.

Newman, N. C., Ryan, P., Lemasters, G., Levin, L., Bernstein, D., Hershey, G. K., Lockey, J. E., Villareal, M., Reponen, T., Grinshpun, S., Sucharew, H., and Dietrich, K. N. 2013. Traffic-related air pollution exposure in the first year of life and behavioral scores at 7 years of age. *Environ Health Perspect*, 121, 731–6.

Peterson, B. S., Rauh, V. A., Bansal, R., Hao, X., Toth, Z., Nati, G., Walsh, K., Miller, R. L., Arias, F., Semanek, D., and Perera, F. 2015. Effects of prenatal exposure to air pollutants (polycyclic aromatic hydrocarbons) on the development of brain white matter, cognition, and behavior in later childhood. *JAMA Psychiatry*, 72, 531–40.

Power, M. C., Kioumourtzoglou, M. A., Hart, J. E., Okereke, O. I., Laden, F., and Weisskopf, M. G. 2015. The relation between past exposure to fine particulate air pollution and prevalent anxiety: Observational cohort study. *BMJ*, 350, h1111.

Power, M. C., Weisskopf, M. G., Alexeeff, S. E., Coull, B. A., Spiro, A. III., and Schwartz, J. 2011. Traffic-related air pollution and cognitive function in a cohort of older men. *Environ Health Perspect*, 119, 682–7.

Raz, R., Roberts, A. L., Lyall, K., Hart, J. E., Just, A. C., Laden, F., and Weisskopf, M. G. 2015. Autism spectrum disorder and particulate matter air pollution before, during, and after pregnancy: A nested case-control analysis within the Nurses' Health Study II Cohort. *Environ Health Perspect*, 123, 264–70.

Schikowski, T., Vossoughi, M., Vierkotter, A., Schulte, T., Teichert, T., Sugiri, D., Fehsel, K., Tzivian, L., Bae, I. S., Ranft, U., Hoffmann, B., Probst-Hensch, N., Herder, C.,

Kramer, U., and Luckhaus, C. 2015. Association of air pollution with cognitive functions and its modification by APOE gene variants in elderly women. *Environ Res*, 142, 10–16.

Szyszkowicz, M., Rowe, B. H., and Colman, I. 2009. Air pollution and daily emergency department visits for depression. *Int J Occup Med Environ Health*, 22, 355–62.

Szyszkowicz, M., Willey, J. B., Grafstein, E., Rowe, B. H., and Colman, I. 2010. Air pollution and emergency department visits for suicide attempts in Vancouver, Canada. *Environ Health Insights*, 4, 79–86.

Thomson, E. M., Vladisavljevic, D., Mohottalage, S., Kumarathasan, P., and Vincent, R. 2013. Mapping acute systemic effects of inhaled particulate matter and ozone: Multiorgan gene expression and glucocorticoid activity. *Toxicol Sci*, 135, 169–81.

Tonne, C., Elbaz, A., Beevers, S., and Singh-Manoux, A. 2014. Traffic-related air pollution in relation to cognitive function in older adults. *Epidemiology*, 25, 674–81.

Volk, H. E., Lurmann, F., Penfold, B., Hertz-Picciotto, I., and Mcconnell, R. 2013. Traffic-related air pollution, particulate matter, and autism. *JAMA Psychiatry*, 70, 71–7.

Wang, Y., Eliot, M. N., Koutrakis, P., Gryparis, A., Schwartz, J. D., Coull, B. A., Mittleman, M. A., Milberg, W. P., Lipsitz, L. A., and Wellenius, G. A. 2014. Ambient air pollution and depressive symptoms in older adults: Results from the MOBILIZE Boston study. *Environ Health Perspect*, 122, 553–8.

Weuve, J., Puett, R. C., Schwartz, J., Yanosky, J. D., Laden, F., and Grodstein, F. 2012. Exposure to particulate air pollution and cognitive decline in older women. *Arch Intern Med*, 172, 219–27.

Wilker, E. H., Preis, S. R., Beiser, A. S., Wolf, P. A., Au, R., Kloog, I., Li, W., Schwartz, J., Koutrakis, P., Decarli, C., Seshadri, S., and Mittleman, M. A. 2015. Long-term exposure to fine particulate matter, residential proximity to major roads and measures of brain structure. *Stroke*, 46, 1161–6.

World Health Organization. 2005. Mental health declaration for Europe: Facing the challenges, building solutions. In *WHO Regional Office For Europe* (ed.). Copenhagen: World Health Organization.

World Health Organization. 2008. *The Global Burden of Disease: 2004 Update*. Geneva, Switzerland: WHO Press.

World Health Organization. 2009. Suicide risk high for young people [Online]. Available: http://www.who.int/mediacentre/multimedia/podcasts/2009/suicide_prevention_20090915/en/

World Health Organization. 2012. Depression: A global crisis. In *WHO Department of Mental Health and Substance Abuse* (ed.). Depression: A Global Public Health Concern. Geneva: World Health Organization.

World Health Organization and Alzheimer's Disease International. 2012. *Dementia: A Public Health Priority*. Geneva: World Health Organization.

Yang, A. C., Tsai, S. J., and Huang, N. E. 2011. Decomposing the association of completed suicide with air pollution, weather, and unemployment data at different time scales. *J Affect Disord*, 129, 275–81.

8 The Impact of China's Vehicle Emission Regulations on Regional Air Quality and Welfare in 2020

Eri Saikawa and Noelle E. Selin

CONTENTS

8.1 INTRODUCTION

Particulate matter (PM) and tropospheric ozone (O_3) influence air quality and cause adverse impacts on human health (Anenberg et al., 2009; Liu et al., 2009; Schwartz et al., 2008; Levy et al., 2001; Dockery et al., 1993). Vehicles are one of the major sources that contribute to emissions of PM and O_3 precursors. China has been experiencing high concentrations of these species, and one of the main reasons is its exponential increase in the number of vehicles. Substantial air pollutant emissions in China come from the road transport sector (Zhang et al., 2009; Ohara et al., 2007).

The number of total vehicles (cars, buses, and trucks) in China increased from 1.78 million in 1980 to 62.8 million in 2009. The number of gasoline vehicles expected in 2020 is 22 times more than that in 2000 (Saikawa et al., 2011). This explosive growth has led to substantial degradation of air quality, especially in urban areas (Cai and Xie, 2007). As a measure to reduce vehicle emissions and enhance air quality, the Chinese central government implemented the first European vehicle emission standards (Euro standards) in 2001, and the standards have been tightened since then (Saikawa, 2013). China nationally implemented the Euro 3 vehicle emission standards in 2008, and some cities such as Beijing and Shanghai have already implemented even more stringent Euro 4 emission standards (Saikawa and Urpelainen, 2014). Although there are further plans to tighten emission standards nationally as well, it is of great interest to quantify the possible impact of China's perfect implementation of the Euro 3 standards for all vehicles (except motorcycles and rural vehicles) by 2020. This is because the actual implementation of these standards will be the key for reducing air pollution in China and within the broader Asian region.

Previous studies have calculated the impact of China's overall air quality. Johnson et al. (1997) estimated the costs of air pollution to be 4.6% of China's GDP (gross domestic product) in 1995. The World Bank and China's State Environmental Protection Administration (2007) together estimated that 1.3% of China's GDP is lost due to air pollution in 2003 in a conservative estimate. Using the willingness-to-pay (WTP) measures, the value increased to 3.8% of China's GDP. Matus et al. (2011) found 5% welfare losses from air pollution–related economic damage in China in 2005. Shindell et al. (2011) estimated the reductions in premature mortalities due to the implementation of the Euro 6 standards in China compared with the current standards to be more than 100,000 deaths in China. These values are significantly larger than the values estimated for the United States (Selin et al. 2009), indicating the magnitude of the problem.

Johnson et al. (1997) and the World Bank study (2007) have quantified health impacts due to PM less than a diameter of 10 μm or less (PM_{10}) exposure only, and these impacts were calculated based on the assumptions of atmospheric concentrations within China due to the lack of data. Selin et al. (2009), Matus et al. (2011), and Shindell et al. (2011) have used global chemical transport model results to estimate exposure, but the grid resolution was rather coarse, with 4° latitude by 5° longitude for the former two, and 2° latitude by 2.5° longitude for the latter, possibly unable to capture the finer nonlinear O_3 formation mechanism. Here, we quantify the health benefits from the reduced surface O_3 and PM smaller than a diameter of 2.5 μm or less ($PM_{2.5}$)

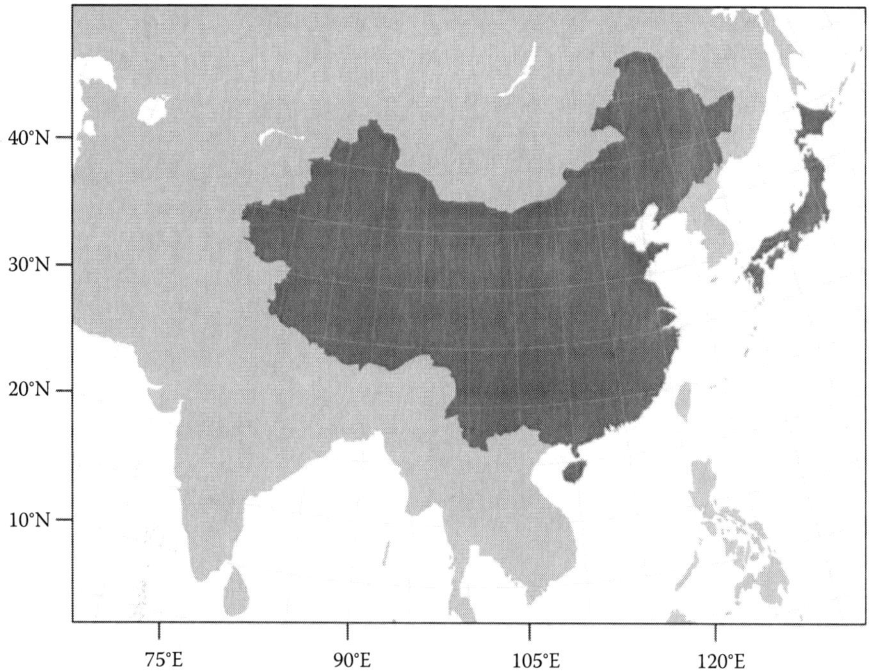

FIGURE 8.1 WRF/Chem model domain.

and the welfare gains in 2020 in China and Japan, due only to China's implementation of the Euro 3 vehicle emission standards compared with no regulations within China.

We estimate atmospheric O_3 and $PM_{2.5}$ at a fine 40 km horizontal resolution using Weather Research and Forecasting coupled with Chemistry (WRF/Chem) for the domain that includes China and Japan (see Figure 8.1 for the model domain). We do simulations for the two scenarios: (1) implementing no vehicle emissions regulation in China (NoPol) and (2) perfect implementation of the Euro 3 vehicle emission standards in China (Euro 3) (Saikawa et al., 2011). We use a general equilibrium (GE) MIT Emissions Prediction and Policy Analysis model with health effects (EPPA-HE) module to assess the increase in social welfare both in China and Japan due to the implementation of vehicle emission standards in China by 2020.

8.2 METHODS

8.2.1 Overall Analysis

Using emissions for the two scenarios—(1) where we assume no vehicle emission standards in China from 2000 to 2020 (NoPol) and (2) where we assume all vehicles except motorcycles and rural vehicles gradually meet the Euro 3 emission standards by 2020 (Euro 3)—we estimate the atmospheric surface O_3 and $PM_{2.5}$ in China and Japan in 2020 with WRF/Chem, as described in Saikawa et al. (2011). We then calculate population-weighted O_3 and $PM_{2.5}$ for China and Japan, and use these as inputs to a GE model to quantify health impacts and economic costs related to the exposure to these species. We also conduct uncertainty analysis based on the confidence intervals of the concentration–response functions to estimate the range in the health effects estimate and economic calculations.

8.2.2 Regional Chemical Transport Model

8.2.2.1 Model Description

We use the regional on-line chemical transport model WRF/Chem (Grell et al. 2005 and references therein). WRF/Chem is a state-of-the-art model where the transport and chemistry is calculated at the same time step. It has been used widely, used for modeling the United States, Mexico, and Asia (e.g., Grell et al., 2005; Ying et al., 2009; Lin et al., 2010; Wang et al., 2010; Saikawa et al. 2011). We use the horizontal resolution of 40 km × 40 km in this study with 31 vertical levels from the surface to 50 mb. The fine resolution of the model has an advantage to resolve local O_3 formation better than the coarse resolution global models.

8.2.2.2 Emissions

For all areas in the domain where the Regional Emissions in Asia (REAS) is available, we use the REAS emissions policy failure case (PFC) scenario for 2020. Where the REAS emissions do not exist, we use the A2 scenario in the Intergovernmental Panel on Climate Change (IPCC) Special Report on Emissions Scenarios (SRES), which assumes the constant economic growth as is the case in PFC. For the road transport sector in China, we use the modified emissions from the REAS inventory to create two 2020 simulations as explained in Saikawa et al. (2011). In the NoPol scenario, we assume that emission factors for carbon monoxide (CO), nitrogen oxides (NO_x), nonmethane volatile organic compound (NMVOC), black carbon (BC), and organic carbon (OC) stay the same as those in 2000 before the implementation of the Euro 1 (the least stringent) vehicle emission standards. In the Euro 3 scenario, we assume that all vehicles meet the Euro 3 standards in 2020, except motorcycles and rural vehicles, as the former is not currently regulated and because it is unclear how well the latter will be regulated. The emissions for the two scenarios are the same except for China's road transport sector. Detailed explanations of the emissions are explained in Saikawa et al. (2011).

8.2.2.3 Model Simulation

We run WRF/Chem for 4 months—January, April, July, and October—to capture seasonality while minimizing the computer use. For each 1-month run, we spin up for 2 weeks. These runs give us the monthly average for each scenario, from which we create the annual mean O_3 mole fractions and $PM_{2.5}$ concentrations by taking the average of these 4 months. One simulation is done for 2000 (baseline) and two simulations are done for 2020—NoPol and Euro 3. These two 2020 scenarios allow us to quantify the health impacts due to surface O_3 and $PM_{2.5}$ changes only by China's implementation of the Euro 3 vehicle standards, compared with no regulations. As an input to the GE model described below, we calculate population-weighted concentrations. We first quantify annual average O_3 mole fractions and $PM_{2.5}$ concentrations for China and Japan in 2000 and 2020 using our WRF/Chem model results. Next, we calculate the population-weighted mixing ratio/concentration for O_3/$PM_{2.5}$ in 2000, using the gridded population distribution estimated at the Center for International Earth Science Information Network (CIESIN, 2005) for 2000. For quantifying the 2020 value, we use the 2015 CIESIN population grid and multiply it by the population change expected in 2020 relative to 2015, using the United Nations population estimate (United Nations, 2010).

8.2.3 Human Health and Economic Model

8.2.3.1 General Description

We use the MIT EPPA-HE model (Paltsev et al. 2005; Matus et al. 2008; Nam et al. 2010). EPPA is a multiregion, multisector computable general equilibrium (CGE) model, but we use it here to quantify the impacts in China and Japan only. EPPA-HE has been used for several policy analyses to quantify the costs and air pollution in different regions (Matus et al. 2008; Nam et al. 2010; Selin et al. 2009).

The model uses the population-weighted concentrations in 16 regions (other than China and Japan, there are 14 regions that we do not analyze and thus stay as constant in the model) as inputs, and calculates cases of health-related diseases and premature mortalities as well as costs to the economy due to lost labor, services and leisure time. When there are health incidents, resources need to be devoted to health care and that prevents the money to be used for the rest of the economy. Labor and leisure time lost due to illness or death are valued at prevailing wage rates. Matus et al. (2008) describes further details, including the economic assumptions of the EPPA-HE model.

8.2.3.2 Concentration–Response Relationship

Table 8.1 describes the concentration–response functions used in the EPPA-HE model to quantify the adverse health effects from the atmospheric O_3 mole fractions and $PM_{2.5}$ concentrations; 95% confidence intervals associated with these relations as found in epidemiology studies are also listed. We assume that the concentration–response relationship is the same in China and Japan, but we realize that this may not be correct. Later, we conduct a sensitivity analysis to account for the uncertainty that might be induced by this assumption as well as the relationship themselves. All the concentration–response relationships are assumed to be linear without a threshold, as suggested in studies for O_3 (e.g. Bell et al., 2006) and for $PM_{2.5}$ (e.g. Schwartz et al., 2008). We calculate the reduced number of morbidity and mortality from acute exposure due to China's implementation of the Euro 3 emission standards for each species and for Japan and China, separately by the following equation:

$$\Delta \text{Cases}_{ijk} = \text{CR}_{ijk} \times \Delta \text{C}_{jk} \times \text{P}_k$$

where $\Delta \text{Cases}_{ijk}$, CR_{ijk}, ΔC_{jk}, P_k indicate the difference in the number of cases for health effect i and pollutant j between the two scenarios (NoPol and Euro 3) in

TABLE 8.1
Concentration–Response Functions for O_3 (Unit of cases $year^{-1}$ $person^{-1}$ ppb^{-1}) and $PM_{2.5}$ (Unit of cases $year^{-1}$ $person^{-1}$ $\mu g\ m^{-3}$), and Adverse Health Costs for Japan and China (USD 2000)

Adverse Health Impacts	O_3 Concentration–Response (95%CI)	$PM_{2.5}$ Concentration–Response (95%CI)	Cost China	Cost Japan
Respiratory hospital admissions	6.25×10^{-6} (-2.5×10^{-6}, 1.50×10^{-5})	1.17×10^{-5} (6.38×10^{-6}, 5.15×10^{-6})	3300	300
Cerebrovascular hospital admissions		8.40×10^{-6} (6.46×10^{-7}, 1.62×10^{-5})	3300	300
Cardiovascular hospital admissions		7.23×10^{-6} (3.62×10^{-6}, 1.09×10^{-5})	3300	460
Symptom days	1.65×10^{-2} (2.85×10^{-3}, 3.15×10^{-2})		60	1
Symptom days (adults)		0.22 (0.025, 0.41)	60	1
Symptom days (children)		0.31 (0.15, 0.46)	60	1
Acute mortality[a]	0.015% (0.005%, 0.02%)	1% (0.66%, 1.33%)	53800	710
Chronic bronchitis (adults)		4.42×10^{-5} (-3.12×10^{-6}, 9.02×10^{-5})	312000	8600
Chronic bronchitis (children)		2.68×10^{-3} (2.07×10^{-4}, 5.17×10^{-3})	600	14
Chronic cough (child only)		3.45×10^{-3} (2.65×10^{-4}, 6.12×10^{-3})	60	1
Restricted activity days (adults)		9.02×10^{-2} (7.92×10^{-2}, 0.10)	200	2
Minor restricted day (adults)	5.75×10^{-3}	5.77×10^{-2}	60	1

(Continued)

TABLE 8.1 (*Continued*)

Concentration–Response Functions for O_3 (Unit of cases year^{-1} person^{-1} ppb^{-1}) and $PM_{2.5}$ (Unit of cases year^{-1} person^{-1} μg m^{-3}), and Adverse Health Costs for Japan and China (USD 2000)

	O_3	$PM_{2.5}$	Cost	
Adverse Health Impacts	**Concentration–Response (95%CI)**	**Concentration–Response (95%CI)**	**China**	**Japan**
	(2.20×10^{-3}, 9.30×10^{-3})	(4.68×10^{-2}, 6.86×10^{-2})		
Congestive heart failure (elderly)		3.08×10^{-5}	20000	300
		(2.37×10^{-6}, 5.93×10^{-5})		
Asthma attack (all asthma)	2.15×10^{-3}	0.15	90	4
	(1.65×10^{-4}, 4.15×10^{-3})	(−0.15, 0.46)		
Bronchodilator (asthma adult)	3.65×10^{-2}	3.00×10^{-2}	2	0
	(-1.28×10^{-2}, 0.0785)	(−0.12, 0.18)		
Bronchodilator (asthma child)		0.28	2	0
Wheeze (asthma child)	8.00×10^{-3}	0.22	18	1
	(-2.15×10^{-2}, 4.05×10^{-2})	(0.025, 0.41)		
Lower respiratory symptoms (adults)		0.31	120	4
		(0.15, 0.46)		
Lower respiratory symptoms (children)		2.92×10^{-5}	120	4

Source: Bickel, P. and Friedrich, R., *Extern E–Externalities of Energy: Methodology 2005 Update*, European Commission, Luxembourg, 2005; Holland, M., Berry, J., and Forster, D., *ExternE, Externalities of Energy* vol 7, Metholodology European Commission, Directorate-General XII, Science Research and Development 1998 Update, Luxembourg, 1999; Holland, M., Hurley, F., Hunt, A., and Watkiss, P., *Methodology for the Cost-Benefit Analysis for CAFE vol 3 Uncertainty in the CAFE CBA: Methods and First Analysis*, AEA Technology Environment, Didcot, 2005; Matus, K., Nam, K. M., Selin, N. E., Lamsal, L. N., Reilly, J. M., and Paltsev, S., *Global Environmental Change*, 22(1):55–66, 2012.

[a] Units are % increase in annual mortality year^{-1} person^{-1} ppb^{-1} for O_3 and annual mortality year^{-1} person^{-1} ppb^{-1} for $PM_{2.5}$.

country k; concentration–response relationship for health effect i and pollutant j in country k; the difference in the air pollutant concentration between the two scenarios (NoPol and Euro 3) for pollutant j; and affected population group in country k, respectively.

We calculate acute mortality in a similar manner with the following equation, using the baseline mortality rate, which we take from the World Bank Global Burden of Disease Study (Lopez et al., 2006):

$$\Delta\text{Cases}_{kt}^{\text{AM}} = \sum_{j} (\text{CR}_{jkt}^{\text{AM}} \times \Delta\text{C}_{jkt} \times M_{kt}^{\text{All}} \times \text{P}_{kt})$$

where $\Delta\text{Cases}_{kt}^{\text{AM}}$, $\text{CR}_{jkt}^{\text{AM}}$, ΔC_{jkt}, $\text{M}_{kt}^{\text{All}}$, and P_{kt} indicate the difference in the number of acute mortality between the two scenarios (NoPol and Euro 3) in country k at time t; acute mortality concentration–response relationship for pollutant j in country k at time t; the difference in concentrations between the two scenarios (NoPol and Euro 3) for pollutant j in country k at time t; the baseline mortality rate for country k at time t; and affected population group in country k at time t, respectively. We use the baseline mortality rate of 0.86% for Japan and the rate of 0.94% for China.

We also calculate mortality due to chronic exposure to $PM_{2.5}$ for adults separately, following the methodology by Matus et al. (2011). Based on the recommendation by Bickel and Friedrich (2005), we assume that chronic mortality only occurs in population groups of age 30 or older. We create five separate age-cohort groups (30–44, 45–59, 60–69, 70–79, and 80+), and calculate concentration–response function for chronic mortality in each group using the following equation:

$$\text{CR}_{kl}^{\text{CM}} = \text{CR}_{k}^{\text{CM}} \times \frac{\text{M}_{kl}^{\text{CPL}} / \text{M}_{kl}^{\text{All}}}{\text{M}_{k}^{\text{CPL}} / \text{M}_{k}^{\text{All}}}$$

where CR^{CM} and M^{CPL} refer to concentration–response function for chronic mortality; and mortality rates for cardiopulmonary diseases, respectively. The subscript k is for each country and l for each cohort group. We use cardiopulmonary disease here, because the long-term exposure to $PM_{2.5}$ is known to be associated with its increase in premature mortality (Pope et al., 2002).

Using the age-conditioned CR functions and the following equation, we calculate the number of mortality due to chronic exposure to $PM_{2.5}$:

$$\Delta\text{Cases}_{kl}^{\text{CM}} = \text{CR}_{kl}^{\text{CM}} \times \left(\frac{1}{2020 - a_l} \times \sum_{i=a_l}^{2020} \Delta\text{C}_{ik} \right) \times \text{M}_{kl}^{\text{All}} \times \text{P}_{kl}$$

where a_l and ΔC_{ik} indicate average birth year for a cohort group l, and $PM_{2.5}$ concentration difference between the two scenarios in time i in country k. All the values used for this calculation are provided in Table 8.2. In order to compute ΔC_{ik}, we make an assumption that we can linearly interpolate the population-weighted concentration

TABLE 8.2

Age, Average Birth Year, Mortality Rates for Cardiopulmonary Diseases, All Mortality Rates, and 2020 Population for Five Cohort Groups in China and Japan Used in EPPA-HE

			China			Japan		
Cohort Group	Age	Avg. Birth Year	CPL Mortality	All Mortality	2020 Population	CPL Mortality	All Mortality	2020 Population
1	30–44	1983	0.0007	0.0043	296,000,000	0.0003	0.0014	26,000,000
2	45–59	1968	0.0038	0.0100	172,000,000	0.0013	0.0045	20,000,000
3	60–69	1955	0.0150	0.0270	66,900,000	0.0050	0.0130	9,820,000
4	70–79	1945	0.0413	0.0626	34,200,000	0.0162	0.0332	6,410,000
5	80+	1935	0.1133	0.1520	9,850,000	0.0683	0.1095	2,840,000

values between 2000 and 2020 for both of the scenarios. We consider the uncertainty associated with this in Section 8.2.3.4.

8.2.3.3 Economic Valuation for Health Impacts

Table 8.1 shows the valuation of each adverse health effects used in the model to calculate economic costs. In order to partition total costs to total service demand, labor lost and leisure lost, we assume the ratio of the money spent in each of these three fields for every health impact. The given valuation values are calculated based on the incurred costs of treating a specific illness in the country (e.g., hospital visits) and also on survey data on WTP to avoid these adverse health impacts. Economic costs for service, labor, and leisure are calculated linearly by using the number of cases for each adverse health impacts, costs associated with them, and the ratio for service, labor and leisure, respectively. The cost estimates for China are taken from Matus (2005), and those for Japan are interpolated from the European values by using PPP. For mortality from acute exposure, we make an assumption that each exposure counts as a 0.5 years of life lost, and calculate a value by applying a statistical life year (VOLY) as recommended by Bickel and Friedrich (2005); 0.5 years is chosen as epidemiological study has shown that most vulnerable population are those that are affected by these acute exposures (Bickel and Friedrich, 2005). We also calculate the lost labor and lost leisure time from mortality due to chronic exposure, based on the years that were lost from an expected 75-year life.

8.2.3.4 Uncertainty Analysis

To account for uncertainties in health impacts and welfare calculation, we conduct uncertainty analysis for both. We use a probabilistic approach with Monte Carlo sampling method, as described in Selin et al. (2009) and Webstser et al. (2012). For economic valuation of health, we assume that the values are always higher in Japan than in China. Using the uncertainty range shown in Table 8.1, we construct probability distributions of concentration–response and costs associated with adverse health impacts, with an assumption that concentration–response functions are correlated at $r = 0.9$. We calculate the uncertainty range using EPPA-HE, based on 400 sets of inputs for each scenario with varying concentration–response functions and with varying costs.

8.3 RESULTS

8.3.1 POPULATION-WEIGHTED CONCENTRATIONS

Figure 8.2 shows the spatial distribution of the annual difference between 2020 NoPol and Euro 3 surface O_3 mixing ratios and $PM_{2.5}$ concentrations. These indicate that the change in mixing ratio/concentration is most visible in the urban areas, where the vehicle number is the largest. These areas are also populated, and thus we see a large difference in numbers between area-weighted and population-weighted mixing ratios/concentrations, especially for $PM_{2.5}$ as described in Table 8.3. For example, whereas the difference between 2020 NoPol and Euro 3 in area-weighted annual mean mixing ratio for O_3 is 3.87 ppbv within China, that for population-weighted annual mean is 7.51 ppbv, almost double the value. The same is true for $PM_{2.5}$ concentrations: the area-weighted value is 2.27 μg m^{-3}, whereas the population-weighted value is 5.85 μg m^{-3}. What is interesting, however, is the fact that we do not see the same impact within Japan. There is no difference between the two for O_3, and there is even smaller difference for $PM_{2.5}$, using the population-weighted average. This is because pollution is transported from China to Japan to places where population is not necessarily large for $PM_{2.5}$, whereas O_3 is transported equally to all of Japan. The difference in lifetime (approximately 2 weeks for the former and 1 month for the latter) results in these spatial distributions.

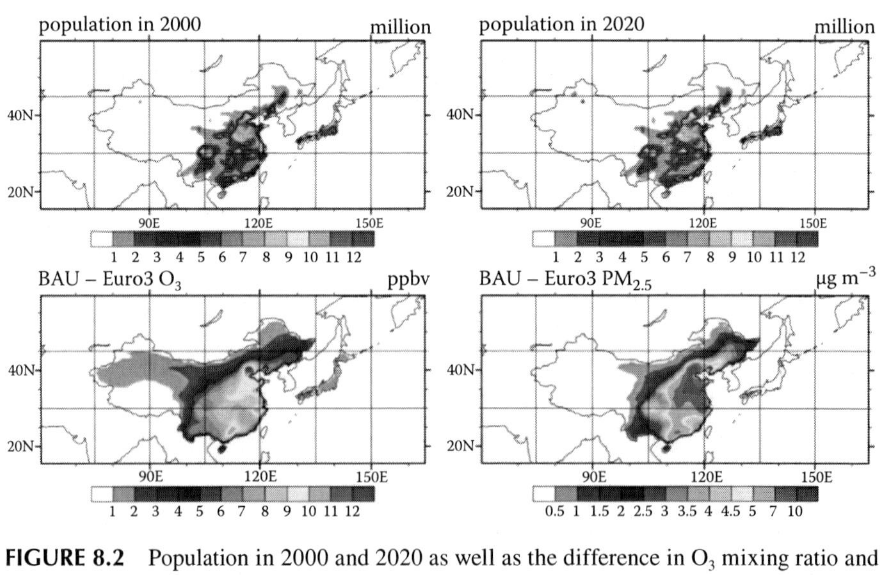

FIGURE 8.2 Population in 2000 and 2020 as well as the difference in O_3 mixing ratio and $PM_{2.5}$ concentration from the two scenarios in 2020.

TABLE 8.3
Area-Weighted and Population-Weighted O_3 Mixing Ratios and $PM_{2.5}$ Concentrations in China and Japan for 2000, for the Two Scenarios in 2020, and the Difference between the Two 2020 Scenarios

	China		Japan	
	O_3 (ppbv)	$PM_{2.5}$ (μg m^{-3})	O_3 (ppbv)	$PM_{2.5}$ (μg m^{-3})
	2000			
Area-weighted	41.06	5.56	44.67	5.27
Population-weighted	42.18	13.59	48.21	6.53
	2020 NoPol			
Area-weighted	51.81	10.89	50.21	6.81
Population-weighted	61.40	27.16	53.91	8.02
	2020 Euro 3			
Area-weighted	47.94	8.62	48.65	6.40
Population-weighted	53.89	21.31	52.35	7.63
	2020 NoPol - Euro 3			
Area-weighted	3.87	2.27	1.56	0.41
Population-weighted	7.51	5.85	1.56	0.39

8.3.2 Reduction of Adverse Health Effects due to China's Vehicle Emissions Regulation

Table 8.4 shows the reduced number of adverse health impacts by implementing the Euro 3 emission standards in China relative to no regulation. We present the values separately for each species for China and Japan, and find that there is a significantly larger number of reduced adverse health impacts in China, but we also find that the number in Japan is not negligible. Furthermore, we quantify the number of mortalities due to chronic exposure to $PM_{2.5}$. Whereas EPPA-HE estimates over 42,700 people to be saved by the implementation of Euro 3 in China, it calculates 190 to be saved in Japan as well. Including both acute and chronic mortalities from surface O_3 and $PM_{2.5}$, EPPA-HE projects a reduction of 154,000 mortalities in China and 1,490 mortalities in Japan simply by China's implementation of the Euro 3 vehicle emission standards.

8.3.3 Welfare Increase due to China's Vehicle Emissions Regulation

Table 8.5 shows the welfare increase due to the reduction of surface O_3 and $PM_{2.5}$ in China and Japan, due to China's implementation of the Euro 3 vehicle emission standards. Using EPPA-HE, we estimate China's and Japan's welfare increases of \$33.4 and \$3.9 billion, respectively, in 2020 due simply to China's implementation of the Euro 3 vehicle emission standards. Similar to the health impacts, whereas 83% of

TABLE 8.4
Reduced Adverse Health Impacts (2020 NoPol–2020 Euro 3) in China and Japan

	China		Japan	
	O_3	$PM_{2.5}$	O_3	$PM_{2.5}$
Respiratory hospital admissions	275,000	61,500	4,900	300
Cerebrovascular hospital admissions	0	44,100	0	300
Cardiovascular hospital admissions	0	38,000	0	200
Symptom days	725,000,000	0	13,000,000	0
Symptom days (adults)	0	228,000,000	0	1,450,000
Symptom days (children)	0	355,000,000	0	1,560,000
Acute mortality	62,000	49,300	1,000	300
Chronic bronchitis (adult)	0	100,700	0	700
Chronic bronchitis (children)	0	5,420,000	0	23,800
Chronic cough (child only)	0	6,970,000	0	30,700
Restricted activity day (adults)	0	317,000,000	0	2,000,000
Minor restricted activity day (adults)	155,000,000	0	3,180,000	0
Congestive heart failure (elderly)	0	8,600	0	103
Asthma attacks (all asthma)	3,770,000	0	67,600	0
Bronchodilator (asthma adult)	48,300,000	24,000,000	967,000	152,000
Bronchodilator (asthma child)	0	13,000,000	0	56,400
Wheeze (asthma child)	76,800,000	0	1,070,000	0
Lower respiratory symptoms adult	0	228,000,000	0	1,450,000
Lower respiratory symptoms child	0	355,000,000	0	1,560,000
Mortality from chronic exposure to $PM_{2.5}$		42,700		190

TABLE 8.5
Welfare Increase in Japan and China in 2020 due to China's Implementation of the Euro 3 Vehicle Emission Standards (in billion USD 2000)

	Japan	China
From reduced $PM_{2.5}$	2.06	27.60
From reduced O_3	1.80	5.77
Total welfare gain	3.87	33.37
2010 GDP %	0.07	1.02

the welfare gain in China originates in the reduction of $PM_{2.5}$ concentrations, a little over 53% is due to the reduction of surface $PM_{2.5}$ in Japan, and the rest due to the reduction of surface O_3. We compare this welfare gain to 2010 GDP values in USD 2000, and find that it shares more than 1% of 2010 GDP for China. The value is small for Japan, and it is less than 0.1%, but we still find it to be important, as this is merely a positive spillover effect from China to Japan. We analyze the relative cost of these values as Japan's environmental aid in the Discussions section.

8.4 DISCUSSIONS

8.4.1 Comparison of Adverse Health Impacts in China and Japan

Whereas mortalities due to $PM_{2.5}$ exposure share 60% of all mortalities in China, they share only 31% in Japan. This is because O_3 has a longer lifetime (~1 month) compared with $PM_{2.5}$ (~2 weeks), and thus the impact of O_3 mixing ratio is seen more vividly in Japan than that of $PM_{2.5}$, as can be found in Figure 8.2. There are larger effects in Japan due to surface O_3, because of reduced transport of O_3 itself from China as well as the reduction in O_3 precursor emissions in China leading to less O_3 formation in Japan. Despite the larger adverse health impacts of $PM_{2.5}$ in comparison with O_3, there are larger health benefits due to reduced O_3 mixing ratios within Japan. This analysis therefore confirms that the reduction of vehicle emissions has a larger impact than reducing its local pollution, as it also reduces regional pollution as a result of decreased CO, NO_x, and VOC emissions as well as PM.

Comparing the economic impact, we find that Japan receives approximately 10% of China's welfare gain as a result of the positive spillover from China's vehicle emissions regulations. Due to the difference in their economic levels, the economic gain is only 0.07% of Japan's GDP as opposed to China's welfare gain by more than 1% of its GDP. Although Japan's welfare gain appears small, we argue that this is substantial, especially considering that this is equivalent to the amount of Japanese ODA to China every year. Japan is the largest donor to China in the world, and since the start of its ODA program in 1979 until 2005, Japan has implemented approximately 3.1 trillion yen in loan aid, 145.7 billion yen in grant aid, and 144.6 billion yen in technical cooperation to China alone (MOFA, 2005). Japan's GDP in 2005 was 5,367,620 billion yen; 0.07% of the GDP value is 3.757 trillion yen, and this is in the same order of magnitude as the total ODA that has been donated from Japan to China over the 26 years. In the recent years, Japan's focus has been on providing environmental aid when giving ODA to China, and our analysis explains why Japan might be most interested in that.

8.4.2 Comparison with Other Literature

Our estimates of the environmental costs for China lie within the range of the previous estimates. In 2007, the recent report by the World Bank and State Environmental Protection Administration (2007) estimated that China's air pollution led to 1.3% of its GDP loss in 2003, in a conservative estimate. Others found a much higher rate, including Matus et al. (2011) that argued for 5% welfare losses from air pollution in China in 2005. We find that China potentially gains approximately 1% of its GDP simply by regulating vehicles, and it is not too surprising, considering the large share of vehicles as a source of CO, NO_x, and VOC emissions.

Shindell et al. (2011) estimated the reductions in premature mortalities due to the implementation of Euro 6 standards in China compared with the current standards to be more than 100,000 deaths in China. We find that China avoids 42,700 chronic mortality and over 111,000 acute mortality by the implementation of Euro 3 standards. This is easily explained by the very little difference in the reduction of air pollutant concentration from Euro 3 to 6 standards. Saikawa et al. (2011) argue that holding everything else constant, the perfect implementation of the Euro 6 emission standards in China reduce CO, NMVOC, BC, and OC emissions only by an additional 1.6, 0.57, 3.0, and 0.68%, respectively, compared with that of the Euro 3 standards. This results in the similar number of premature mortality in the two studies.

8.5 CONCLUSIONS AND POLICY IMPLICATIONS

We quantified the reduced adverse health impacts and the increased welfare in China and Japan due to China's implementation of the Euro 3 vehicle emission standards in 2020. We used a fine-resolution WRF/Chem chemical transport model to analyze the change in surface air quality and the EPPA-HE model to calculate the impacts. We found that 154,000 mortalities in China and 1,490 mortalities in Japan are saved simply by China's implementation of the Euro 3 vehicle emission standards, compared with no regulations in 2020. Furthermore, this policy implementation in China leads to the welfare increase of \$33.4 and \$3.9 billion in China and Japan, respectively. We quantified that China increases its welfare by more than 1% of its 2010 GDP by the implementation of the Euro 3 vehicle emission standards. Japan's welfare gain is 0.07% of its 2010 GDP, but this is also significant, considering that it is merely a positive spillover effect from China.

We have assumed that emissions increase in other sectors except for the road transport sector in China, and we did not include emissions regulations in other areas. Vehicle emissions are unique in that we see reductions in both surface O_3 and $PM_{2.5}$ simultaneously. We found that China benefits more from the reduction of

$PM_{2.5}$ due to its large adverse health impacts, but Japan benefitted equally from the two species, because of a longer lifetime of O_3 and other O_3 precursor species that were reduced in China. Our analysis reconfirms that China's air pollution regulations potentially have large health benefits not only in China but also in Japan and other surrounding countries.

REFERENCES

Anenberg, S. C., West, J. J., Fiore, A. M., Jaffe, D. A., Prather, M. J., Bergmann, D., Cuvelier, K., Dentener, F. J., Duncan, B. N.,Gauss, M., Hess, P., Jonson, J. E., Lupu, A., Mackenzie, I. A., Marmer, E., Park, R. J., Sanderson, M. G., Schultz, M., Shindell, D. T., Szopa, S., Vivanco, M. G., Wild, O., and Zeng, G. 2009. Intercontinental impacts of ozone pollution on human mortality. *Environ Sci Technol*, 43, 6482–7.

Bell, M. L., Peng, R. D., and Dominici, F. 2006. The exposure-response curve for ozone and risk of mortality and the adequacy of current ozone regulations. *Environ Health Perspect*, 114, 532–6.

Bickel, P. and Friedrich, R. 2005. *Extern E–Externalities of Energy: Methodology 2005 Update*. Luxembourg: European Commission.

Cai, H. and Xie, S. 2007. Estimation of vehicular emission inventories in China from 1980 to 2005. *Atmos Environ*, 41, 8963–79.

CIESIN (2005). Socioeconomic Data and Applications Center (SEDAC) http://sedac.ciesin.columbia.edu/gpw/global.jsp

Dockery, D. W., Pope, C. A., Xu, X., Spengler, J. D., Ware, J. H., Fay, M. E., Ferris, B. G., and Speizer, F. E. 1993. An association between air pollution and mortality in six US cities. *N Engl J Med*, 329, 1753–9.

Grell, G. A., Peckham, S. E., Schmitz, R., Mckeen, S. A., Frost, G., Skamarock, W. C., and Eder, B. 2005. Fully coupled "online" chemistry with the WRF model. *Atmos Environ*, 39, 6957–75.

Holland, M., Berry, J., and Forster, D. 1999. *ExternE, Externalities of Energy* vol 7, Metholodology European Commission, Directorate-General XII, Science Research and Development 1998 Update. Luxembourg.

Holland, M., Hurley, F., Hunt, A., and Watkiss, P. 2005. *Methodology for the Cost-Benefit Analysis for CAFE vol 3 Uncertainty in the CAFE CBA: Methods and First Analysis*, AEA Technology Environment, Didcot.

Johnson, T. M.; Liu, F., and Newfarmer, R. 1997. *Clear Water, Blue Skies: China's Environment in The New Century*. Washington, DC: The World Bank. http://documents.worldbank.org/curated/en/1997/09/694613/clear-water-blue-skies-chinas-environment-new-century

Levy, J. I., Carrothers, T. J., Tuomisto, J. T., Hammitt, J. K., and Evans, J. S. 2001. Assessing the public health benefits of reduced ozone concentrations. *Environ Health Perspect*, 109, 1215–26.

Lin, M., Holloway, T., Carmichael, G. R., and Fiore, A. M. 2010. Quantifying pollution inflow and outflow over East Asia in spring with regional and global models. *Atmos Chem Phys*, 10, 4221–39.

Liu, J., Mauzerall, D. L., and Horowitz, L. W. 2009. Evaluating intercontinental transport of fine aerosols: (2) Global health impacts. *Atmos Environ*, 43, 4339–47.

Lopez, A. D., Mathers, C. D., Ezzati, M., Jamison, D. T., and Murray, C. J. L. (ed). 2006. *Global Burden of Disease and Risk Factors*. New York: Oxford University Press.

Matus, K. 2005. Health impacts from urban air pollution in China: The burden to the economy and the benefits of policy. MS Thesis. Cambridge, MA: Engineering Systems Division, Massachusetts Institute of Technology.

Matus, K., Nam, K.-M., Selin, N. E., Lamsal, L. N., Reilly, J. M., and Paltsev, S. 2011. Health damages from air pollution in China. MIT Joint Program Tech Report 196.

Matus, K., Nam, K. M., Selin, N. E., Lamsal, L. N., Reilly, J. M., and Paltsev, S. 2012. Health damages from air pollution in China. *Global Environmental Change*, 22(1):55–66.

Matus, K., Yang, T., Paltsev, S., Reilly, J., and Nam, K.-M. 2008. Toward integrated assessment of environmental change: Air pollution health effects in the USA. *Clim Change*, 88, 59–92.

Ministry of Foreign Affairs (Japan). 2005. Overview of official development assistance (ODA) to China. Available at http://www.mofa.go.jp/policy/oda/region/e_asia/china/index.html. Retrieved on September 22, 2011.

Nam, K. M., Selin, N. E., Reilly, J. M., and Paltsev, S. 2010. Measuring welfare loss caused by air pollution in Europe: A CGE Analysis. *Energy Policy*, 38(9):5059–5071, doi: 10.1016/j.enpol.2010.04.034

Ohara, T., Akimoto, H., Kurokawa, J., Horii, N., Yamaji, K., Yan, X., and Hayasaka, T. 2007. An Asian emission inventory of anthropogenic emission sources for the period 1980 2020. *Atmos Chem Phys*, 7, 4419–44.

Paltsev, S., Reilly, J. M., Jacoby, H. D., Eckaus, R. S., McFarland, J., Sarofim, M., Asadoorian, M., and Babiker, M. 2005. *The MIT Emissions Prediction and Policy Analysis (EPPA) Model: Version 4 MIT Joint Program on the Science and Policy of Global Change Report Series Report 125*. Cambridge, MA: MIT.

Pope, C. A., Burnett, R. T., Thun, M. J., Calle, E. E., Krewski, D., Ito, K., and Thurston, G. D. 2002. Lung cancer, cardiopulmonary mortality, and long-term exposure to fine particulate air pollution. *JAMA*, 287, 1132–41.

Saikawa, E. 2013. *Domestic Politics and Environmental Standards: China's Policy-Making Process for Regulating Vehicle Emissions*, J. Sato (ed.), "Governance of natural resources: Uncovering the social purpose of materials in nature", Chapter 3, pp. 74–97. The United Nations University Press.

Saikawa, E., Kurokawa, J., Takigawa, M., Borken-Kleefeld, J., Mauzerall, D. L., Horowitz, L. W., and Ohara, T. 2011. China's vehicle emissions on regional air quality in 2020: A scenario analysis. *Atmos Chem Phys*, 11, 9465–84.

Saikawa, E. and Urpelainen, J. 2014. Environmental standards as a strategy of international technology transfer. *Environmental Science and Policy*, 38, 192–206.

Schwartz, J., Coull, B. Laden, F., and Ryan, L. 2008. The effect of dose and timing of dose on the association between airborne particles and survival. *Environ Health Perspect*, 116, 64–69.

Selin, N. E., Wu, S., Nam, K. M., Reilly, J. M., Paltsev, S., Prinn, R. G., and Webster, M. D. 2009. Global health and economic impacts of future ozone pollution. *Environ Res Lett*, 4, 1–9.

Shindell, D., Faluvegi, G., Walsh, M., Anenberg, S. C., Dingenen, R. V., Muller, N. Z., Austin, J., Koch, D., and Milly, G. 2011. Climate, health, agricultural and economic impacts of tighter vehicle emission standards. *Nature Climate Change*, 1, 59–66.

United Nations. 2010. Population Division of the Department of Economic and Social Affairs of the United Nations Secretariat. World population prospects: The 2010 revision. Available: http://esa.un.org/unpd/wpp/index.htm.

Wang, X., Liang, X.-Z., Jiang, W., Tao, Z., Wang, J. X.-L., Liu, H., Han, Z., Liu, S., Zhang, Y., Grell, G. A., and Peckham, S. E. 2010. WRF-Chem simulation of East Asian air quality: Sensitivity to temporal and vertical emissions distributions. *Atmos Environ*, 44, 660–9.

Webster, M., Sokolov, A., Reilly, J., Forest, C., Paltsev, S., Schlosser, A., Wang, C., Kicklighter, D., Sarofim, M., Melillo, J., Prinn, R., and Jacoby, H. 2012. Analysis of climate policy targets under uncertainty. *Climatic Change*, 112, 3, 569–583, http://EconPapers.repec.org/RePEc:spr:climat:v:112:y:2012:i:3:p:569–583.

World Bank and State Environmental Protection Administration. 2007. *Cost of Pollution in China: Economic Estimates of Physical Damages*. Washington, DC: World Bank. http://documents.worldbank.org/curated/en/2007/02/7503894/cost-pollution-china-economic-estimates-physical-damages

Ying, Z., Tie, X., and Li, G. 2009. Sensitivity of ozone concentrations to diurnal variations of surface emissions in Mexico city: A WRF/Chem modeling study. *Atmos Environ*, 43, 851–9.

Zhang, Q., Streets, D. G., Carmichael, G. R., He, K. B., Huo, H., Kannari, A., Klimont, Z., Park, I. S., Reddy, S., Fu, J. S., Chen, D., Duan, L., Lei, Y., Wang, L. T., and Yao, Z. L. 2009. Asian emissions in 2006 for the NASA INTEX-B mission. *Atmos Chem Phys*, 9, 5131–53.

9 Health Effects of Coal Energy Generation*

Susan Buchanan, Erica Burt, and Peter Orris

CONTENTS

9.1 INTRODUCTION

Access to electricity has a positive effect on the health and well being of people worldwide (Wang, 2002). The use of coal to generate energy, however, has negative health consequences (Smith et al., 2013). Evidence suggests an impact on health at every stage in its use for electricity generation—from mining to postcombustion disposal (Smith et al., 2013). The combustion of coal has been well studied, with compelling evidence of widespread health effects on the population. Air pollution produced by coal power plants can affect the respiratory and cardiovascular systems, and coal used for heating and cooking indoors generates pollutants known to cause

* This chapter is substantially based, with permission, on Susan Buchanan, Erica Burt, and Peter Orris, "Beyond black lung: Scientific evidence of health effects from coal use in electricity generation," 2014 Macmillan Publishers Ltd. 0197-5897, *Journal of Public Health Policy* Vol. 35, 3, 266–277, doi:10.1057/jphp.2014.16;www.palgrave-journals.com/jphp/.

respiratory ailments and cancer. Coal combustion also contributes to climate change, which can harm human health on a global scale.

We review scientific evidence of health effects from the use of coal for electricity generation, focusing primarily on air emissions from coal combustion. We compile recent evidence of health effects from coal mining, transport, and combustion from the peer-reviewed literature as well as from governmental, international agency, and research institute reports. We searched biomedical research databases (Ovid Medline and PubMed) for articles using the search terms *coal* or *solid fuel* and *health* or *burden* or *economic* or *cost*. We included only English-language articles published in the past 10 years. We also reviewed articles mentioned in the references and included them if appropriate. We gave priority to articles examining coal use in power plants. We generally excluded studies of exposures produced by alternative uses of coal.

9.2 BACKGROUND

When coal is burned in power plants it produces air-borne pollutants: particulate matter (PM), sulfur dioxide (SO_2), oxides of nitrogen (NO_X), carbon dioxide (CO_2), mercury, arsenic, chromium, nickel, other heavy metals, acid gases (HCL, HF), polycyclic aromatic hydrocarbons (PAHs), and varying amounts of radioactive uranium and thorium in fly ash particles.

In 2011, the World Health Organization (WHO) compiled air quality data from 1,100 cities in 91 countries and found that residents living in many urban areas are exposed to persistently elevated levels of fine particle pollution. The report states that in both developed and developing countries, the largest contributors to urban outdoor air pollution include motor transport, small-scale manufacturers and other industries, burning of biomass and coal for cooking and heating, as well as coal-fired power plants (WHO, 2011b).

Forty percent of the electricity produced in the world is generated from burning coal, and the number of power plants burning coal is likely to rise in the next two decades as energy demand increases worldwide (International Energy Agency, 2007, 2012). The World Resources Institute estimates that, globally, 1200 new power plants have been proposed, with 76% of them in China and India (Yang and Cui, 2012). Most of coal's health burden results from burning it in power plants. The remainder results from other steps of coal's life cycle—extraction, transport, and disposal (Dones et al., 2005; Rabl and Spadaro, 2006).

9.3 HEALTH EFFECTS OF COAL-FIRED POWER PLANTS

In their 2007 article in *The Lancet*, Markandya and Wilkinson summarized the health burden of electricity generation using coal and lignite (the softest and most polluting form of coal). The authors estimate that for every Terawatt-hour (TWh) of electricity produced from coal in Europe, there are 24.5 deaths; 225 serious illnesses, including congestive heart failure and chronic bronchitis; and 13,288 minor illnesses (Markandya and Wilkinson, 2007). When lignite is used, each

TWh of electricity produced results in 32.6 deaths, 298 serious illnesses, and 17,676 minor illnesses (Markandya and Wilkinson, 2007). The International Energy Agency reports that worldwide coal-based energy production was 8572 TWh in 2010 (International Energy Agency, 2012). On the basis of per TWh estimates by Markandya and Wilkinson, the worldwide health toll from air pollution due to coal combustion is about 210,000 deaths, almost 2 million serious illnesses, and over 151 million minor illnesses per year. These estimates do not include the effects of climate change.

This calculation of health burden used European pollution levels and population density. In countries with weaker air pollution controls, higher use of coal, lower quality coal, or higher population density close to power plants, the health burden is greater. A study in China, reported by Markandya and Wilkinson in 2007 (Markandya and Wilkinson, 2007), estimated 77 deaths per TWh from a coal-fired power plant that met Chinese environmental standards—more than three times the estimate of deaths per TWh for coal combustion in Europe. For all of China, this would result in an estimated 250,000 deaths per year, based on estimates of coal combustion in China (International Energy Agency, 2012).

9.3.1 Respiratory Effects

Specific pollutants from burning coal harm the respiratory system: PM, SO_2, and NO_X, and others. Injury to the airways and lungs via oxidative stress leads to inflammation, cytotoxicity (direct harm to cells), and cell death.

9.3.1.1 Particulate Matter

Particulates generated by burning coal are characterized by size—particles up to 10 micrometers called *PM_{10}*, and smaller particles less than 2.5 micrometers (*$PM_{2.5}$*), a subset of PM_{10}. $PM_{2.5}$ travels deeper into airways and is believed to cause more harm. A study of many power plants in China found that of the total mass of PM emitted, PM_{10} comprised 62%–84% and $PM_{2.5}$ comprised 8%–44% (Yi et al., 2006).

In a report evaluating over 40 studies on the health effects of exposure to small PM ($PM_{2.5}$), the US Environmental Protection Agency (USEPA) concluded that $PM_{2.5}$ likely causes respiratory symptoms, the development of asthma, and decrements in lung function in children (USEPA, 2009b). The EPA concludes that a 10 $\mu g/m^3$ increase in $PM_{2.5}$ is associated with a 1%–3.4% decrease in forced expiratory volume in 1 sec (FEV1) (USEPA, 2009b). EPA also concluded that exposure to $PM_{2.5}$ increases emergency department visits and hospital admissions for respiratory-related symptoms such as infections and chronic obstructive pulmonary disease. Epidemiological evidence from Australia and New Zealand (Barnett et al., 2005), Mexico (Barraza-Villarreal et al., 2008), Canada (Chen et al., 2004), and Europe (de Hartog et al., 2003) confirms that these effects on the respiratory system are seen wherever communities are exposed to $PM_{2.5}$. In addition to respiratory illnesses, current evidence suggests that long-term exposure to $PM_{2.5}$ is causally linked to the development of lung cancer (USEPA, 2009b).

9.3.1.2 Sulfur Dioxide

Exposure to SO_2 emitted by coal-burning power plants increases the incidence and severity of respiratory symptoms of those living nearby, particularly children with asthma. For adults and children who are susceptible, inhalation of SO_2 causes inflammation and hyperresponsiveness of the airways, aggravates bronchitis, and decreases lung function (USEPA, 2008b). Community-level SO_2 concentration is associated with hospitalizations for asthma and other respiratory conditions, as well as emergency department visits for asthma, particularly among children and adults over 65 years (USEPA, 2008b). A review of epidemiological studies in cities in Italy, Spain, France, and the Netherlands found that low concentrations of SO_2 (less than 10 ppb 24-hour average) are associated with increased risk of death from heart and lung conditions (USEPA, 2008b). For every 10 ppb increase in SO_2 concentration there is a 0.4%–2% increased risk of death (USEPA, 2008b). Fortunately, ambient concentrations of SO_2 in many countries have declined over the last few decades, owing to installation of pollution control technologies at coal-burning power plants. Countries with weaker pollution standards put their populations at risk of SO_2 health effects. The ambient concentrations of SO_2 in China, for example, increased from 2000 to 2006 at an annual rate of 7.3%, mainly because of emissions from power plants. But in 2005, new policy in China increased the use of flue-gas desulfurization technologies, and SO_2 concentrations have begun to decline (Lu et al., 2010).

9.3.1.3 Oxides of Nitrogen

NO_X are by-products of fossil fuel combustion in automobiles and coal-fired power plants, and other places. NO_X react with chemicals in the atmosphere to create pollution products such as ozone in smog, nitrous oxide (N_2O), and nitrogen dioxide (NO_2). NO_2 and ozone are of particular concern. When asthmatic children are exposed to NO_2, they can experience increased wheezing and coughing (USEPA, 2008a). At low concentrations (0.2–0.5 ppm), NO_2 has been found to result in lung function decrements in asthmatics (USEPA, 2008a). Exposure to NO_2 also increases susceptibility to viral and bacterial infections, and at high concentrations (1–2 ppm) can cause airway inflammation (USEPA, 2008a). Increases in ambient NO_2 levels (3–50 ppb) are linked to increases in hospital admissions and emergency department visits for respiratory problems, particularly asthma (USEPA, 2008a).

9.3.2 Cardiovascular Effects

Coal-fired power plants contribute to the global burden of cardiovascular disease primarily through the emission of PM. $PM_{2.5}$ has been causally linked to cardiovascular disease and death (USEPA, 2009b). The WHO estimates that worldwide, 5% of cardio-pulmonary deaths are due to PM pollution (WHO, 2013). The mechanism of cardiovascular injury is the same as for the respiratory system: vascular oxidative stress leads to vessel inflammation and cytotoxicity. Long-term exposure to $PM_{2.5}$ has been shown to accelerate the development of atherosclerosis and increase emergency department visits and hospital admissions for ischemic heart disease and congestive heart failure. The USEPA reports that a majority of the studies it reviewed found a 0.5%–2.4% increase in emergency department visits and hospital admissions for cardiovascular diseases per each 10 μg/m^3 increase in $PM_{2.5}$ concentrations (USEPA, 2009b). A 2007 scientific review of the health effects of combustion emissions reported an 8%–18% increase in cardiovascular deaths per 10 μg/m^3 increase in $PM_{2.5}$ concentration in the United States (Lewtas, 2007). Recent studies conducted in China and Latin America confirm the significant link between outdoor air pollution and cardiovascular events (Liu et al., 2013; Romieu et al., 2012).

9.3.3 Reproductive Effects

Research has documented that exposure to air pollution during pregnancy can cause low birth weight (Sram et al., 2005). Studies that investigated the effects of SO_2 and PM (China, South Korea), and NO_2, CO, and ozone (South Korea), concluded that air pollution containing these constituents was associated with low birth weight (Sram et al., 2005). In studies evaluating the association between electricity generation at coal-fired power plants and infant mortality, infant mortality was shown to have increased with increased use of coal in countries that had mid to low infant mortality rates at baseline (1965), such as Chile, China, Mexico, Thailand, Germany, and Australia, although this effect was not seen in those countries with high baseline infant mortality rates (Gohlke et al., 2011).

9.3.4 Neurologic Effects

9.3.4.1 Mercury

When coal is burned, mercury vapor is released into the atmosphere. The United Nations estimates that 26% of global mercury emissions (339–657 metric tons/year) come from burning coal in power plants (Pacyna et al., 2010). The mercury from coal-burning power plants is deposited into waterways, converted to methyl-mercury, and passed up the aquatic food chain (Lippmann et al., 2003; National Research Council (US), 2010). Local, regional, and distant mercury emissions contaminate fish. Methyl-mercury-contaminated fish, when eaten by pregnant women, can cause developmental effects in their offspring, such as delayed neurodevelopment, plus subtle changes in vision, memory, and language (WHO, 2007). Epidemiological studies suggest that many newborns and children around the world have levels of mercury in their bodies that put them at risk of these adverse effects. Data from the United States suggest that more than 300,000 newborns each year are born at risk for these effects (Mahaffey et al., 2004). A study in Spain found 42% of the preschool or newborn children tested had mercury levels in their hair above the EPA reference concentration for safety, 1 μg Hg/g. A study in Hong Kong estimates that a majority of children exceed safety levels of mercury because of consumption of mercury-contaminated fish (Diez et al., 2009; Lam et al., 2013).

9.3.5 Life Expectancy

A study modeling the effect on life expectancy of coal power generation predicted a decrease in life expectancy in countries with moderate life expectancy at the baseline year (1965), including Poland, China, Mexico, and Thailand. In India and

China, years of life lost were estimated to be up to 2.5 years and 3.5 years, respectively (Gohlke et al., 2011).

9.4 CLIMATE CHANGE

Global climate change is caused by the accumulation of greenhouse gases in the Earth's atmosphere. Two of the major greenhouse gases contributing to climate change are products of coal combustion: CO_2 and N_2O. As the concentrations of these gases in the atmosphere increase, the average global temperature slowly increases, setting in motion a host of consequences that further promote climate change such as melting of polar ice and thawing of arctic permafrost.

As the average global temperature increases, researchers predict public health will suffer, particularly in low-income countries that have fewer resources to respond and adapt to the changes brought on by warmer global temperatures (Costello et al., 2009). A higher average global temperature and warmer oceans are already increasing the occurrence of extreme weather events such as floods, hurricanes, and droughts that, in turn, increase disease and injury and adversely affect water quality and food security (USEPA, 2009a, 2009c). Warmer average temperatures alter ecosystems, decreasing some key food-chain supporting species such as corals, and increasing the growing ranges of some weeds, grasses, and trees that may further increase the severity and prevalence of allergies (USEPA, 2009a, 2009c). Other consequences include the spread of climate-sensitive diseases such as tick- and mosquito-borne diseases and of food- and waterborne pathogens; an increase in ground-level ozone and smog, which aggravate asthmas and increase hospital visits; and an increase in the number of extremely hot days, which can cause heat-related mortality (Costello et al., 2009; USEPA, 2009a, 2009c; McMichael et al., 2006; Vardoulakis and Heaviside, 2012). The mass migration of people to avoid these climate-related consequences may cause conflict and further stress on water, food, shelter, sanitation, and health care resources (International Energy Agency, 2007).

9.5 HEALTH EFFECTS OF INDOOR COAL COMBUSTION

Using solid fuels such as coal for heating and cooking is estimated to cause 910,000 deaths from acute lower respiratory infections in children under 5 years and 693,000–1 million deaths from chronic obstructive pulmonary disease per year worldwide (WHO, 2011a; Smith et al., 2004). Approximately 0.4 billion people worldwide, many of them in China, use coal to cook and heat their homes. In 2000, the WHO conducted a meta-study on the use of solid fuels for heating and cooking and reported that 12.9% adults in East Asia and 2.1% adults in South Asia are exposed to coal smoke from heating and cooking, causing accelerated loss of lung function and over 16,000 deaths from lung cancer per year (Smith et al., 2004).

9.6 HAZARDS OF COAL EXTRACTION

The occupational health impacts of mining coal are well known and must be considered when reviewing the effects of electricity generation with coal. In a 2002 review of 250 studies on coal mining, Stephens and Ahern (2001) calculated that up to 12% coal miners develop coal workers' pneumoconiosis and silicosis because of the inhalation of dust during mining operations. Miners are also at higher risk for chronic bronchitis and accelerated loss of lung function. Most research on the health effects of coal mining has been undertaken among miners in large mines in Europe and North America (Stephens and Ahern, 2001). Small scale mines, many of which are found in developing countries, are often more hazardous, resulting in higher rates of accidents and injuries. They often employ less-experienced workers and children. These populations have increased vulnerability to occupational disease and injury (Stephens and Ahern, 2001).

9.7 COST OF COAL ENERGY GENERATION

The impacts of burning coal can be described in economic terms, and several papers have attempted to estimate the cost of using coal by assigning value to the environmental and public health damage caused during coal's extraction, transportation, combustion, and disposal. One such study by Epstein et al. (2011) estimated that the external costs of coal-fired electricity production in the United States add an extra USD 0.178 to each kWh of electricity produced; an amount that would triple its cost to consumers. Another US report by Machol et al. (2013) estimated USD 0.19–0.45 per kWh as the cost of the health burden and environmental damages from coal combustion. As part of an analysis for the European Commission in 2005, Rabl et al. estimated the external life-cycle costs of fossil fuels (the most expensive of which was coal) to be 0.016–0.058 €/kWh (Yang and Cui, 2012).

In 2011, the USEPA estimated the benefits and costs of the Clean Air Act, a law that regulates emissions of SO_2, NO_X, carbon monoxide, and PM in the United States. The EPA calculated that the ratio of health care cost savings to compliance costs was 25:1 in 2010, meaning that for every dollar spent complying with the Clean Air Act, 25 dollars were saved in welfare, ecological, and health care costs owing to lower disease burden, including a reduction in premature deaths, bronchitis, asthma, and myocardial infarction (USEPA, 2011).

REFERENCES

Barnett, A. G., Williams, G. M., Schwartz, J., Neller, A. H., Best, T. L., Petroeschevsky, A. L., and Simpson, R. W. 2005. Air pollution and child respiratory health: A case-crossover study in Australia and New Zealand. *Am J Respir Crit Care Med,* 171, 1272–8.

Barraza-Villarreal, A., Sunyer, J., Hernandez-Cadena, L., Escamilla-Nunez, M. C., Sienra-Monge, J. J., Ramirez-Aguilar, M., Cortez-Lugo, M., Holguin, F., Diaz-Sanchez, D., Olin, A. C., and Romieu, I. 2008. Air pollution, airway inflammation, and lung function in a cohort study of Mexico City school children. *Environ Health Perspect,* 116, 832–8.

Chen, Y., Yang, Q., Krewski, D., Shi, Y., Burnett, R. T., and Mcgrail, K. 2004. Influence of relatively low level of particulate air pollution on hospitalization for COPD in elderly people. *Inhal Toxicol,* 16, 21–5.

Costello, A., Abbas, M., Allen, A., Ball, S., Bell, S., Bellamy, R., Friel, S., Groce, N., Johnson, A., Kett, M., Lee, M., Levy, C., Maslin, M., Mccoy, D., Mcguire, B., Montgomery, H., Napier, D., Pagel, C., Patel, J., de Oliveira, J. A., Redclift, N., Rees, H., Rogger, D., Scott, J., Stephenson, J., Twigg, J., Wolff, J., and Patterson, C. 2009. Managing the health effects of climate change: Lancet and University College London Institute for Global Health Commission. *Lancet,* 373, 1693–733.

de Hartog, J. J., Hoek, G., Peters, A., Timonen, K. L., Ibald-Mulli, A., Brunekreef, B., Heinrich, J., Tiittanen, P., Van Wijnen, J. H., Kreyling, W., Kulmala, M., and Pekkanen, J. 2003. Effects of fine and ultrafine particles on cardiorespiratory symptoms in elderly subjects with coronary heart disease: The ULTRA study. *Am J Epidemiol,* 157, 613–23.

Diez, S., Delgado, S., Aguilera, I., Astray, J., Perez-Gomez, B., Torrent, M., Sunyer, J., and Bayona, J. M. 2009. Prenatal and early childhood exposure to mercury and methylmercury in Spain, a high-fish-consumer country. *Arch Environ Contam Toxicol,* 56, 615–22.

Dones, R., Heck, T., Bauer, C., Hirschberg, S., Bickel, P., Preiss, P., Panis, L. I., and Vlieger, I. D. 2005. *Externalities of Energy: Extension of Accounting Framework and Policy Applications*. Stuttgart, Germany: Paul Scherrer Institute.

Epstein, P. R., Buonocore, J. J., Eckerle, K., Hendryx, M., Stout III, B. M., Heinberg, R., Clapp, R. W., May, B., Reinhart, N. L., Ahern, M. M., Doshi, S. K., and Glustrom, L. 2011. Full cost accounting for the life cycle of coal. *Ann N Y Acad Sci,* 1219, 73–98.

Gohlke, J. M., Thomas, R., Woodward, A., Campbell-Lendrum, D., Pruss-Ustun, A., Hales, S., and Portier, C. J. 2011. Estimating the global public health implications of electricity and coal consumption. *Environ Health Perspect,* 119, 821–6.

International Energy Agency. 2007. *World Energy Outlook 2007: China and India Insights*. Paris, France: Organization for Economic Co-operation and Development (OECD).

International Energy Agency. 2012. *Key World Energy Statistics 2012*. Paris, France: International Energy Agency.

Lam, H. S., Kwok, K. M., Chan, P. H., So, H. K., Li, A. M., Ng, P. C., and Fok, T. F. 2013. Long term neurocognitive impact of low dose prenatal methylmercury exposure in Hong Kong. *Environ Int,* 54, 59–64.

Lewtas, J. 2007. Air pollution combustion emissions: Characterization of causative agents and mechanisms associated with cancer, reproductive, and cardiovascular effects. *Mutat Res,* 636, 95–133.

Lippmann, M., Cohen, B., and Schlesinger, R. 2003. *Environmental Health Science: Recognition, Evaluation, and Control of Chemical and Physical Health Hazards*. New York: Oxford University Press.

Liu, L., Breitner, S., Schneider, A., Cyrys, J., Bruske, I., Franck, U., Schlink, U., Marian Leitte, A., Herbarth, O., Wiedensohler, A., Wehner, B., Pan, X., Wichmann, H. E., and Peters, A. 2013. Size-fractioned particulate air pollution and cardiovascular emergency room visits in Beijing, China. *Environ Res,* 121, 52–63.

Lu, Z., Streets, D. G., Zhang, Q., Wang, S., Carmichael, G. R., Cheng, Y. F., Wei, C., Chin, M., Diehl, T., and Tan, Q. 2010. Sulfur dioxide emissions in China and sulfur trends in East Asia since 2000. *Atmos Chem Phys,* 10, 6311–31.

Machol, B., and Rizk, S. 2013. Economic value of U.S. fossil fuel electricity health impacts. *Environ Int,* 52, 75–80.

Mahaffey, K. R., Clickner, R. P., and Bodurow, C. C. 2004. Blood organic mercury and dietary mercury intake: National Health and Nutrition Examination Survey, 1999 and 2000. *Environ Health Perspect,* 112, 562–70.

Markandya, A., and Wilkinson, P. 2007. Electricity generation and health. *Lancet,* 370, 979–90.

McMichael, A. J., Woodruff, R. E., and Hales, S. 2006. Climate change and human health: Present and future risks. *Lancet,* 367, 859–69.

National Research Council (US). 2010. *Hidden Costs of Energy: Unpriced Consequences of Energy Production and Use,* Washington, DC, National Academy Press.

Pacyna, J., Sundseth, K., Pacyna, E., and Panasiuk, N. D. 2010. Study on Mercury Sources and Emissions and Analysis of Cost and Effectiveness of Control Measures: 'UNEP Paragraph 29 Study'. United Nations Environment Programme.

Rabl, A. and Spadaro, J. V. 2006. Environmental impacts and costs of energy. *Ann N Y Acad Sci,* 1076, 516–26.

Romieu, I., Gouveia, N., Cifuentes, L. A., De Leon, A. P., Junger, W., Vera, J., Strappa, V., Hurtado-Diaz, M., Miranda-Soberanis, V., Rojas-Bracho, L., Carbajal-Arroyo, L., Tzintzun-Cervantes, G., and Committee, H. E. I. H. R. 2012. Multicity study of air pollution and mortality in Latin America (the ESCALA study). *Res Rep Health Eff Inst*, 171, 5–86.

Smith, K. R., Frumkin, H., Balakrishnan, K., Butler, C. D., Chafe, Z. A., Fairlie, I., Kinney, P., Kjellstrom, T., Mauzerall, D. L., Mckone, T. E., Mcmichael, A. J., and Schneider, M. 2013. Energy and human health. *Annu Rev Public Health,* 34, 159–88.

Smith, K. R., Mehta, S., and Maeusezahl-Feuz, M. 2004. Indoor air pollution from household use of solid fuels. *In:* Ezzati, M., Lopez, A. D., Rodgers, A., and Murray, C. J. L. (eds.) *Comparative Quantification of Health Risks: Global and Regional Burden of Disease Attributable to Selected Major Risk Factors.* Geneva, Switzerland: World Health Organization.

Sram, R. J., Binkova, B., Dejmek, J., and Bobak, M. 2005. Ambient air pollution and pregnancy outcomes: A review of the literature. *Environ Health Perspect,* 113, 375–82.

Stephens, C. and Ahern, M. 2001. *Worker and Community Health Impacts Related to Mining Operations Internationally: A Rapid Review of the Literature.* London, UK: Mining and Minerals for Sustainable Development.

US Environmental Protection Agency. 2008a. *Integrated Science Assessment for Oxides of Nitrogen—Health Criteria.* Washington, DC: USEPA.

US Environmental Protection Agency. 2008b. *Integrated Science Assessment for Sulfur Oxides—Health Criteria.* Washington, DC: USEPA.

US Environmental Protection Agency. 2009a. *EPA's Endangerment Finding: Health Effects* [Online]. Available: http://www.epa.gov/climatechange/Downloads/endangerment/EndangermentFinding_Health.pdf

US Environmental Protection Agency. 2009b. *Integrated Science Assessment for Particulate Matter.* Washington, DC: USEPA.

US Environmental Protection Agency. 2009c. *Technical Support Document for Endangerment and Cause or Contribute Findings for Greenhouse Gases under Section 202(a) of the Clean Air Act* [Online]. Available: http://www.epa.gov/climatechange/Downloads/endangerment/Endangerment_TSD.pdf

US Environmental Protection Agency. 2011. *The Benefits and Costs of the Clean Air Act: 1990–2020.* Washington DC: EPA Office of Air and Radiation.

Vardoulakis, S. and Heaviside, C. 2012. *Health Effects of Climate Change in the UK 2012, Current Evidence, Recommendations and Research Gaps.* UK: Health Protection Agency.

Wang, L. 2002. Health outcomes in poor countries and policy options: Empirical findings from demographic and health surveys. *Policy Research Working Paper 2831.* The World Bank.

World Health Organization. 2007. *Exposure to Mercury: A Major Public Health Concern. Public Health and Environment.* Geneva, Switzerland: World Health Organization.

World Health Organization. 2011a. *Indoor Air Pollution and Health, Fact sheet N°292* [Online]. Available: http://www.who.int/mediacentre/factsheets/fs292/en/.

World Health Organization. 2011b. *Tackling the Global Clean Air Challenge* [Online]. Available: http://www.who.int/mediacentre/news/releases/2011/air_pollution_20110926/en/ [Accessed 9 May 2013].

World Health Organization. 2013. *Global Health Observatory (GHO): Outdoor Air Pollution* [Online]. Available: http://www.who.int/gho/phe/outdoor_air_pollution/en/index.html.

Yang, A. and Cui, Y. 2012. Global coal risk assessment: Data analysis and market research. *WRI Working Paper.* Washington DC: World Resources Institute.

Yi, H., Hao, J., Duan, L., Li, X., and Guo, X. 2006. Characteristics of inhalable particulate matter concentration and size distribution from power plants in China. *J Air Waste Manag Assoc,* 56, 1243–51.

Index